T0323370

The AI Playbook

Management on the Cutting Edge Series

Abbie Lundberg, series editor

Published in cooperation with *MIT Sloan Management Review*

The AI Playbook

Mastering the Rare Art of
Machine Learning Deployment

Eric Siegel

The MIT Press
Cambridge, Massachusetts
London, England

The MIT Press would like to thank the anonymous peer reviewers who provided comments on drafts of this book. The generous work of academic experts is essential for establishing the authority and quality of our publications. We acknowledge with gratitude the contributions of these otherwise uncredited readers.

This book was set in ITC Stone Serif Std and ITC Stone Sans Std by New Best-set Typesetters Ltd. Printed and bound in the United States of America.

Library of Congress Cataloging-in-Publication Data

Names: Siegel, Eric, 1968– author.
Title: The AI playbook : mastering the rare art of machine learning deployment / Eric Siegel.
Description: Cambridge, Massachusetts : The MIT Press, [2024] | Series: Management on the cutting edge | Includes bibliographical references and index.
Identifiers: LCCN 2023017997 (print) | LCCN 2023017998 (ebook) | ISBN 9780262048903 (hardcover) | ISBN 9780262378130 (epub) | ISBN 9780262378123 (pdf)
Subjects: LCSH: Business—Data processing. | Machine learning.
Classification: LCC HF5548.2 (ebook) | LCC HF5548.2 .S44865 2024 (print) | DDC 658/.05 23/eng/20230—dc08
LC record available at https://lccn.loc.gov/2023017997

10 9 8 7 6 5 4 3 2 1

This book is dedicated with all my heart to my mother,
Lisa Schamberg, and my father, Andrew Siegel.

Contents

This book's notes—references, plus resources for further learning—are available at www.bizML.com.

For a tutorial glossary that includes the terms introduced within this book and more, see www.MachineLearningGlossary.com.

Series Foreword

The world does not lack for management ideas. Thousands of researchers, practitioners, and other experts produce tens of thousands of articles, books, papers, posts, and podcasts each year. But only a scant few promise to truly move the needle on practice, and fewer still dare to reach into the future of what management will become. It is this rare breed of idea—meaningful to practice, grounded in evidence, and *built for the future*—that we seek to present in this series.

Abbie Lundberg
Editor in Chief
MIT Sloan Management Review

Foreword

There's almost no business outcome that machine learning cannot help you improve today. From delivering a best-in-class customer and consumer experience to fueling productivity, increasing safety, optimizing operations, and improving your employee experience, ML can raise the bar on the metrics that matter across all. Its practical deployment represents the forefront of human progress: *improving operations with science*. But where do you start, and how do you ensure what you do start doesn't end up in the dustbin?

Over the course of my career I've consulted with over thirty Fortune Global 500 companies on data and analytics, and led global data and analytics organizations at Caterpillar and Unilever. I've seen the highs and the lows, including analytics programs that generate tremendous value and competitive advantage, and those that never seem to leave the starting gate. In my experience, those companies or teams that struggle to embed analytics at scale typically suffer not because of imperfect analytics execution or ML models, but rather because of a gap in the other factors required for success.

As one example, while consulting, I worked with an analytics team at one of the world's largest retailers on a program to improve marketing ROI. The in-house team had already developed an advanced media analytics model. They were flush with data, leveraging hundreds of millions of data points on marketing spend, response, products, stores, and other contributing factors. The team poured hours and hours into perfecting the model and fine-tuning it to highest possible levels of accuracy and then summarizing the output into a list of top insights for action. The day of the big presentation to marketing leadership arrived and the team presented the recommendations to improve ROI by making key changes to offline marketing spend. They looked to top marketing leadership for their reaction, expecting smiles, gratitude, praise, and appreciation. Instead, they were met with a mix of apathy and disbelief. The problem was that the team had missed crucial steps required to fully understand and incorporate stakeholder priorities, decision-making factors, and processes.

Contrast that with an experience I had leading an AI-powered portfolio optimization program at Unilever. Unilever is a global organization. The products are sold in over 25 million stores across 190 countries, with over 2.5 billion people using the products every day. Unilever's brands include Dove, Knorr, Sunsilk, Hellmann's, Axe, Ben & Jerry's, Domestos, Suave, TRESemmé, and Magnum.

We saw an opportunity to make smarter and faster decisions by taking a global, data-driven approach to optimize our portfolio of products and reduce complexity—through a program we would later name Polaris. A sharper portfolio of products ultimately benefits consumers and retailers, optimizes our operations, and drives profitable growth for Unilever's shareholders. Our team built an AI-powered capability and business process to analyze the entire product portfolio globally and recommend products to delist, grow, fix, and protect. The system leverages analytics to track the execution of those actions and drive accountability across thousands of individuals in the organization. We created and scaled Polaris globally in approximately two years, bringing together the best of machine and human intelligence, which empowered us to make more efficient and effective decisions and grow through simplification.

The path to get there wasn't easy and there wasn't a guidebook available to help us at the time. Fortunately for the reader, the steps outlined in this book bring to life crucial best practices we followed in delivering a globally scaled initiative with lasting business impact. These include:

1. **Start with outcomes in mind and focus on delivering value incrementally.**

   We started with a simple question: Could we increase the rate of decision making and execution to simplify the product portfolio—delivering savings while driving growth with our customers? Only after delivering on that scope and establishing that value did we expand to complete product portfolio optimization, including non-consumer facing simplification such as flagging specifications and ingredients to harmonize across products.

2. **Leverage empathy to overcome barriers to change.**

   Consciously or unconsciously, we are all preprogrammed to resist change. To overcome this, the analytics team spent hundreds of hours with other teams across the business to understand how portfolio decisions were being taken currently—including marketing, sales, supply chain, finance, research and development, and retailers. By gaining an understanding of the pain points in the current processes, we were able to bring forward a compelling value proposition for stakeholders across levels and functions.

3. **Prepare the data so that it meets business needs.**

   Only by anticipating early the differences in data availability, due to the global nature of our business, did the team succeed in scaling the capability across geographies. We recognized that we had to adapt to variations of data across markets—some of which were rich with retailer and third-party data illuminating shopper behavior patterns, while others held inconsistent point of sale and shopper information based on the route to market. A versatile data infrastructure and stringent data validation process were key to success.

These experiences have made me acutely aware of the many hurdles that must be overcome to deliver scaled value realization with ML. Innovating the enterprise with ML is revolutionary, and revolutions aren't easy.

Many senior data leaders come to learn the same lessons, but only after years of experience and failed projects. Then after understanding it themselves, they still struggle to advocate for these success factors with their business counterparts. Without common understanding between business stakeholders and data leaders on the best practices for delivering data and analytics transformations, many projects fail to take off, struggle to scale, or ultimately don't deliver on the business outcomes.

The industry needs a framework to better leverage ML for business results. This book introduces *bizML*, which brings forward the best practices in a succinct and actionable way. Not only is the book a timely and much needed addition to the industry; it is also powerful in bringing AI down to earth, eschewing the hype, and making it tangible for all readers. This book is the driver's manual for machine learning—every business and analytics professional should read it.

Morgan Vawter
Global Vice President of Data & Analytics at Unilever,
former Chief of Analytics at Caterpillar,
former Data Management Practice Lead at Accenture,
and a *Fortune* magazine "40 Under 40" honoree

Preface

A Brief History of Why Machine Learning Projects Stall

When promoting breakthrough technology, be careful what you wish for.

Back in the Dark Ages, before data was cool and phones were smart, I networked my way into the swank office of a powerful business executive. Hoping that he would introduce me to—or become—my first client, I declared that I was striking out on my own as a machine learning (ML) consultant. Unfamiliar with ML and disinterested, he looked at me like, "Don't waste my time," and I was quickly back on the streets of San Francisco.

This was 2003, right after I'd relocated from the East Coast and ordered new business cards, all in the pursuit of my passion. I had fallen in love with ML a dozen years earlier, first in the research lab and then as a Columbia University professor teaching the graduate-level ML and AI courses. It was the most exciting, potent, and widely applicable kind of technology. Moving west, I vowed to introduce it to the non-academic world. I wanted to see ML deployed.

At that time, a corner of the industrial world was already using ML, but they called it something else: *data mining*. I thought that term was misleading to the non-data folks, but "machine learning" kept getting me kicked out of offices. So I latched onto a new buzzword that had just started to gain traction, *predictive analytics*. A rose by any other name.

Unfortunately, my improved vocabulary didn't immediately land me clients. "You should just take a full-time job," a senior executive at an established analytics vendor bluntly threw in my financially insecure face.

Instead, I doubled down. Tripled down. I held corporate training seminars. I published articles. I networked like mad.

Clients eventually started coming in, but only enough to keep me busy. I was knee-deep in demand, but I needed it up to my belly button. The world still didn't get it. I had to evangelize harder. I took a three-pronged approach:

1. *Conference.* I launched Machine Learning Week (formerly Predictive Analytics World), the first ML conference series outside academic and vendor events. Bolstered by its sister publication, the *Machine Learning Times*, the conference series has since grown to serve 18,000 attendees internationally.

2. *Book.* Next, I wrote *Predictive Analytics*, the first popular book that showed readers of all levels how the algorithms work under the hood. Written to ignite and excite, it ended up becoming a best-seller, winning several awards, landing me 100 keynote speeches at conferences outside my own, and being adopted as course material by hundreds of universities.

3. *Music video.* I even dropped an educational rap music video called "Predict This!," which went a bit viral (to watch, go to www.Predict This.org). Surely this proves that I'd literally do anything to spread the gospel of ML.

Whether or not these efforts helped light the fuse, one thing's for sure: ML exploded in popularity. It grew from a nascent industry to a full-blown commercial movement. It came of age as a core enterprise practice necessary to sustain competitive advantage. Hyperboles reigned as *data scientist* dethroned *firefighter* to become "the sexiest job."

Watching ML become so hot felt both gratifying and surreal. The experience reinforced an age-old lesson: *Keep the faith.* When you believe in a good idea—such as the notion that learning from data is not only cool but valuable—and stick to your convictions, people will eventually come around.

Failure to Launch

Unfortunately, ML's great rise has also taught me another lesson: *Be careful what you wish for.* The buzz has gone too far. In a way, ML is now too hot for its own good. The problem is, the onslaught of excitement has fed a common misconception that derails many ML projects:

> **The ML Fallacy**: Since ML algorithms work (amazing and true), the models they generate are intrinsically valuable (not true).

The value of ML comes only by launching it to enact organizational change. After generating a model with ML, you capture its potential value only when you deploy it so that it actively improves operations. Until a model is *used* to actively reshape how your organization works, it's *use-less*—literally. A model doesn't solve any business problems on its own and it ain't gonna deploy itself. ML can be the disruptive technology it's cracked up to be, but only if you disrupt with it.

Most ML projects fail to deploy. I believe this is mainly because most ML leaders neglect to properly plan for the operational change that deployment would bring to fruition. That planning takes more preaching, socializing, cross-disciplinary collaboration, and change-management panache than many, including myself, initially realized.

Far too often, the data scientist delivers a viable model, but the operational team isn't ready for the pass—and they drop the ball. There are wonderful exceptions and glowing successes, but the generally poor track record we witness today forewarns of broad disillusionment with ML—even a dreaded *AI winter*. It's time to tap the brakes and correct course so that ML can deliver on its promise.

Breaking through the ML Snafu

So I've pivoted from ML cheerleader to wary disciplinarian—albeit an optimistic one—with a new mission: *Standardize and broadcast the very particular business discipline needed to get ML launched.* Whereas my first book was about how ML works technically, this book is about how to run ML projects so that models not only work in the lab but also successfully deploy.

First things first: Business professionals—who are a primary audience for this book—need some edification. Before those in charge can confidently green-light model deployment, they must gain a concrete understanding of how an ML project works from end to end: *What will the model predict? Precisely how will those predictions affect operations? Which metric meaningfully tracks how well it predicts?* and *What kind of data is needed?*

Only when the business leaders—including executives, managers, and decision makers—come up to speed on this semi-technical but straightforward knowledge can we bridge the gap between the tech and business sides and bring model deployment into the realm of possibility.

These days, everything I do is to unite those two worlds, tech and biz. In addition to this book, I've taken another three-pronged approach:

1. *Conferences focused on deployment.* Newer offshoots of my event series, Machine Learning Week, build on the nuts-and-bolts aspect of analytics to also cover industry-specific deployment, including applications in marketing, financial services, industry 4.0, healthcare, and climate technology. The first track devotes itself to the business side—we call it the *operationalization and leadership* track.

2. *Business school professorship.* After a twenty-two-year hiatus, I returned to academia to hone the methodology described in this book, serving for one year as the Bodily Professor in Analytics at the Darden School of Business at the University of Virginia. The switch in departments—from computer science years ago to business more recently—reflects my shift in focus: For ML to succeed, we need a business-side vantage.

3. *More expansive training.* Finally, I've launched an online course, "Machine Learning Leadership and Practice: End-to-End Mastery," to broaden the almost universally narrow focus of today's ML courses—which typically jump straight to the number crunching, forgoing the extensive business planning that should come first.

If you don't have time for a three-month course, you might instead just read this book. It covers the disciplined approach required to deploy ML initiatives, formulated as a six-step playbook that I call *bizML*. Along the way, it gets readers of all backgrounds up to speed on the semi-technical knowledge they need.

Considering the innumerable dollars and resources pumped into ML, how much more potential value could we capture by adopting a universal procedure that facilitates the collaboration and planning needed to reach deployment?

Let's find out.

Optional FAQ: What This Book Is about and Who It's For

You may skip this optional FAQ—but I advise looking through the questions below for those that pertain to or interest you. Readers of this book come with diverse backgrounds and various preconceptions about the problem the book aims to solve: getting machine learning deployed. This FAQ will orient you, clarifying why you should read this book and aligning your expectations.

What is this book about?

This book presents a strategic and tactical playbook for launching machine learning, a six-step discipline to run an ML project so that it successfully deploys. I call this practice *bizML*.

Along the way, the book also delivers the semi-technical background knowledge everyone participating in the project needs—in a friendly, accessible way anyone can understand. Because of that coverage, the book also serves as a non-technical introduction to the field for newcomers.

Why does machine learning need a specialized business practice?

Here's the problem. ML is the world's most powerful generally applicable technology. But ML can only improve large-scale operations by *changing* them. For that reason, an ML project shouldn't be viewed as "a *technology* project." Instead, to make an impact, it must be reframed as a *business* project meant to improve operational performance, with ML as only one component—one that's necessary but not sufficient.

With the attention overwhelmingly focused on the technical portion and its execution, the industry has failed to establish a widely adopted business practice for executing *the whole other half* of a successful ML project. As a result, new ML initiatives routinely fail to deploy.

If most machine learning projects fail to deploy, is the field of machine learning a flop?

Not at all. Many ML projects succeed, even if it's only a minority—even a fraction of this popular field's many projects is still many. Moreover, in certain circumstances an ML project is bound to succeed, such as high-priority projects at a Big Tech firm or projects meant to update an existing model that's already deployed. The industrial world remains bullish on ML because its great potential remains intact.

Since this book covers a practice for running ML projects, is it only for leaders?

No. When an ML project follows bizML, the organizational practice presented by this book, everyone involved in the project participates in that practice in some way. Only with universal familiarity with this end-to-end practice—and with the semi-technical background knowledge that drives it—can the team collaborate most effectively.

Who is this book for?

This book serves anyone who wishes to gain value with ML by participating in its business deployment, no matter whether you'll play a role on the business side or the technical side.

First and foremost, I wrote this book for business professionals—the people who run the ML project, hold stakes in it, make decisions about it, or manage the operations that will be changed (and improved) by it. This includes executives, directors, managers, consultants, and leaders of all kinds.

But this book is for techies, too. If you're a data scientist, ML engineer, or any kind of technical practitioner involved with ML, this book invites you to step back from the hands-on execution and gain a new perspective on the holistic paradigm within which you are contributing.

Is this book a how-to?

This book is a *business* how-to, but not a *technical* how-to. Unlike most ML books, it tackles the business practice instead of the technical practice. It presents a six-step business practice, *bizML*, for running an ML project.

This book does not delve deeply enough to guide data professionals in the technical how-to. That's what the vast majority of other ML books are for. The ML methods they cover are only one ingredient. They constitute a key technical component of the project, but that component makes for only one of the six project steps covered in this book. Accordingly, one and only one chapter of this book, chapter 5, delves into core ML methods—it provides an accessible "crash course."

This book also differs from most *business* books on ML, which present a strategic industry overview. Such books typically cover the topic from a higher level, without providing how-to guidance and without concretely detailing how ML integrates to deliver operational improvements.

What introductory materials should I read before this book?

None are required. This book is accessible for all readers and serves as a conceptually complete introduction to the field of ML for newcomers. While describing the end-to-end steps for executing an ML project, it covers the fundamentals along the way. Gaining some background knowledge of ML methods before reading this book certainly wouldn't hurt, but considering the theme of this book—*the business vantage of an ML project should precede the technical vantage*—you are invited to read this book first and then determine how much further you'd like to dig into the core technology.

I already understand that ML projects must begin with a business objective—do I need this book?

Establishing the deployment goal is only the first step—literally. It's the first of this book's six-step bizML practice. The rest serve to accomplish that goal. Pursuing it demands an in-depth, end-to-end procedure. The helpful mantra "begin with the business objective" alone does not surmount the challenges of deployment. It takes a book.

I'm a business professional, not a data scientist—do I really need semi-technical knowledge?

> Rapid, continuous learning and reskilling . . . starts at the top. AI requires a new type of C-suite leadership, with deep understanding of AI and its implications . . .
> —Julie Sweet, chair and CEO of Accenture

Yes, you must achieve a particular kind of *data literacy* in order to be involved in the deployment of ML, helping to guide each project and ensure that it works within—and successfully produces value for—business operations.

You may be unconvinced. After all, to drive a car, you don't need to know how the engine works. True—but you do need expertise: a keen feel for how the car moves, a sense of the physics, including the momentum of the vehicle and the friction of the tires. As a driver, you've also internalized the rules of the road and you know what moves to expect from other drivers and what they'll expect from you.

Driving an ML project is just the same. To pursue the goal of improving operational performance, you need the what, why, and how much. You need to understand the precise way in which this technology will enact change to business processes, the basis of those changes, and a quantitative appraisal of how well it is working.

Fear not—you don't need a degree in the "rocket science" part, and what you do need to learn goes down easy. It's conceptual, not hands-on, and requires no heavy math. This book delves into the nifty principles of internal combustion, not how to change a spark plug. This level of data literacy is useful for almost everyone, like driver's education, not auto-mechanic school.

I'm a technically trained data professional—why do I need this book?

This book establishes a sorely needed strategic framework, providing complementary business-side know-how that all great data professionals need to master. The real "data science unicorn" isn't the person who knows every analytical technique and technology; rather, it's the one who has expanded their skillset to also participate in a company-wide, business-oriented effort that gets their models deployed. After all, the soft skills are often the hard ones.

In so doing, this book does cover certain *technical* steps generally omitted by courses and books meant for data professionals, including how to fully establish the dependent variable (called the *output variable* in this book), how to prepare the data, and how to establish the performance metric (including why accuracy and a popular technical metric called *AUC* are usually the wrong choice)—all so that these choices align with business objectives and operational considerations.

On the other hand, know that this broadly accessible book is not the technical fare to which you're likely accustomed. For some experienced data professionals, the best use of this book may be to give it a good skim—slowing down to give chapter 0 on the need for a specialized business-side practice and chapter 3 on evaluation metrics a complete read—and then passing it on to your boss or a key colleague.

Is this book about artificial intelligence?

The buzzword *AI* can mean many things, but this book is about ML, which is a central basis for—and what many mean by—*AI*. This book does not cover other areas that are also sometimes referred to as AI, including *artificial general intelligence* (hypothetical systems that would be capable of any intellectual task humans can do), *natural language processing*, *rule-based systems*, and *computer vision*.

Does this book pertain to generative AI?

Yes. *Generative AI* dazzles the world by writing text and producing images—but when it comes to improving operational efficiencies, classical ML (a.k.a. *predictive AI*) has long reigned supreme. However, generative AI is also well suited and stands to potentially beat out classical ML in some arenas. The bizML practice presented by this book also serves generative AI—for projects that apply generative AI to measurably improve great numbers of operational decisions. For either kind of technology, bizML gets you there, guiding the project to a successful deployment.

Does this book pertain to deep learning?

Yes. Although deep learning is more technically complex than many classical ML methods and tends to be applied for different classes of problems (more on image processing, for example, and less on customer prediction), the ML project discipline presented in this book applies and is equally needed. The organizational challenges of deployment are largely the same, no matter how the model being deployed operates on the inside.

Does this book pertain to predictive analytics?

Yes—*predictive analytics* is a major subset of ML. It is the application of ML methods for certain business problems. Alternatively, in many contexts, *predictive analytics* is simply a synonym for *machine learning.*

How does this book compare to your previous book, *Predictive Analytics*?

This book and my previous book—*Predictive Analytics: The Power to Predict Who Will Click, Buy, Lie, or Die*—are complementary, but each stands alone. Neither is required reading for the other, so you can read only one or both, in either order. Both make ML accessible for business professionals, newcomers, and other non-data professionals, but they serve different purposes: *Predictive Analytics* is about how ML works, and this book is about how to capitalize on it.

	The AI Playbook (this book)	*Predictive Analytics* (my previous book)
A business how-to	Yes	–
ML deployment	Yes	The general idea
Performance metrics	Yes	The general idea
Data preparation	Yes	–
Technical modeling methods	A one-chapter overview	**Decision trees, ensembles, uplift modeling—one chapter each**
Technical pitfalls	Misreporting performance	**P-hacking, overfitting, presuming that correlation implies causation**
ML ethics	A brief but wide overview	**A chapter about how ML reveals sensitive information and predictive policing**
Case studies	**UPS, FICO, two dot-coms**	**HP, Chase, NSA, 183 mini-case studies**

To which ML tools and software does this book apply?

This book pertains to all ML software. It is vendor-neutral and tool-agnostic. The contents apply universally, regardless of which of the many ML software tools you or your data professionals may end up using.

Is this book about supervised or unsupervised machine learning?

This book only covers *supervised machine learning*, which trains models over *supervised data*, that is, data that consists of examples for which the target prediction is already known—either by way of accumulating historical outcomes or manually labeling the data (more specifically, the book mostly focuses on *binary classification*, that is, ML for predicting yes/no outcomes). Supervised ML is the kind of ML most commonly applied to optimize business operations. However, the bizML practice presented in this book largely holds for unsupervised learning projects as well.

How is bizML different from MLOps?

The method presented by this book, bizML, is a business practice for running ML projects successfully through to deployment, whereas MLOps is a set of technical methods and practices to manage and maintain models. Although both relate to the operationalization of models, MLOps addresses the *technical execution* of ML projects and bizML addresses the *organizational execution*, including project leadership and cross-functional collaboration. The two work together: A project that follows bizML may well employ MLOps. But no *technical* solution alone can address the *business-side* challenges faced by ML projects. Instead, it takes a business paradigm such as bizML.

How is bizML different from CRISP-DM?

This book's introduction of bizML represents a renewed effort to establish an updated, industry-standard playbook for running successful ML projects that is pertinent and compelling to both business professionals and data professionals. One previous standard established almost thirty years ago, CRISP-DM, paved the way by laying out many of the fundamentals, but never gained much traction among business professionals. For more details, see chapter 0.

Where are this book's notes and glossary?

This book's notes—references, plus resources for further learning—are available at www.bizML.com. For a tutorial glossary that includes the terms introduced within this book and more, see www.Machine LearningGlossary.com.

Introduction

Never sell AI. Instead, pitch operational improvements, with no more than a footnote to mention machine learning as part of the solution.

Most ML leaders focus more on the technology than its deployment, so most new ML initiatives fail.

Make no mistake, operational change is a tough sell, especially in comparison to hot tech, which sells so effortlessly that we actually call it "sexy." It's less glamorous to propose a process overhaul. Folks respond like you're suggesting a root canal. But that's life—great gains come only by imposing great change.

We begin with the tale of an ambitious pioneer determined to go big with ML at a century-old Fortune 500 company already set in its ways. His domain? Logistics. But stay tuned and you'll see why stodgy ML projects are, ironically, really the sexiest. You'll also see how the *Innovator's Paradox* can be overcome and why it is a good thing that, ultimately, most practical ML projects are bound to dispense with the "AI" brand.

They tried to warn him. As Jack Levis pursued a deep-seated desire to innovate, his colleagues thought he was committing career suicide. "I love your passion," one coworker told him, "but you need to know: Everyone thinks you're nuts."

Jack wasn't trying to change the world. He was only taking on the small matter of streamlining how the United Parcel Service (UPS) delivered packages—16 million of them a day. He just couldn't remain satisfied with the status quo. There were 185 million miles of annual driving to potentially shave off.

Jack had tackled this whole crazy idea voluntarily. He hadn't been charged with this project from above, nor was it part of his defined responsibilities at UPS. Instead, he had proactively formed a small team to develop a proof-of-concept prototype. As a group, they'd taken it on part-time, on the side.

One autumn day, after years of work, Jack finally landed an exciting opportunity to present to a key UPS executive named Chuck. So he sat Chuck down and pitched him an inventive story: a system that would prescribe more efficient delivery routes for truck drivers and, in so doing, more fully realize the value of another of Jack's recent contributions, a system that planned for tomorrow's deliveries by predicting them.

Sure, implementing this system would mean introducing a mammoth change to existing operations. But it promised a mammoth payoff. Jack ran through his presentation swiftly and stuck the landing.

But in response, Chuck's face remained blank. After a pause, he cleared his throat and asked, "So, are you working on anything important?"

Jack's heart sank. Years later, he still remembers that day well. "I assure you," he says, "I didn't sleep a wink that night."

The Innovator's Paradox states that the more novel or radical an idea, the greater the struggle to gain support for it. This mighty law seemed to be binding Jack's hands. *How do you sell innovation that's so profound the buyer doesn't grasp it?*

Pioneers Beware: Disrupt at Your Own Peril

Unfortunately, the warning from Jack's coworkers applies far and wide: The greater your innovation's potential impact, the more treacherous it is to pursue. Bravely press onward and there's hell to pay, in the form of doubt, obstinance, resentment, and, perhaps worst of all, a glaring lack of appreciation. In return for your innovation, expect first to be misunderstood and then, eventually, ruthlessly exploited.

The world ostracizes the very innovators who built it. What if you developed the first television (Philo Farnsworth—inspired to emit images row by row like a farmer plows a field) or invented the intermittent windshield wiper (Robert Kearns—more challenging than you think). Say you took off for the first officially observed, sustained flight (Alberto Santos-Dumont—don't let the Wright brothers' "long hop" fool you) or fathered theoretical computer science (Alan Turing—who also cracked Germany's World War II Enigma machine and founded AI philosophy). Things didn't turn out too well for you. Corporate empires came down on Farnsworth and Kearns and governments brutally persecuted Santos-Dumont and Turing. The result? Three deaths and a nervous breakdown.

And yet, in modern times, innovators disrupt more than ever. The greatest opportunities aren't in building a new device such as a TV or airplane. Instead, the leading innovative paradigm upgrades *existing* systems. It infiltrates the established enterprise and overhauls its largest-scale activities, its millions of daily operations. It combats risk, targets advertising, prevents fraud, optimizes manufacturing, triages medical cases, and streamlines logistics.

I'm talking about *machine learning* (ML). This book is about ML in the following practical, applied sense:

> **Machine learning**: Technology that learns from experience (data) to predict the outcome or behavior of each customer, patient, package delivery, business, vehicle, image, piece of equipment, or other individual unit—in order to drive better operational decisions. ML generates a *predictive model* whose job is to calculate a *predictive score* (probability) for each individual.

ML is a central basis for—and what many mean by—*AI*. This book does not cover other areas that are also sometimes referred to as AI, including *artificial general intelligence* (hypothetical systems that would be capable of any intellectual task humans can do), *natural language processing*, *rule-based systems*, and *computer vision*. But this book does pertain to *generative AI*, most famous for writing text and producing images. When it comes to improving operational efficiencies, classical ML has long reigned supreme—but generative AI is also well suited and stands to potentially beat out classical ML in some arenas. The framework presented by this book also serves generative AI—for projects that apply generative AI to measurably improve great numbers of operational decisions.

ML innovates in a straightforward, albeit disruptive way. Don't let the glare emanating from this glitzy technology obscure the simplicity of its fundamental duty: For most business applications, the purpose of ML is to issue actionable predictions—which is why it's also sometimes called *predictive analytics*. Although learning from data in order to generate a predictive model deserves as much "gee-whiz" admiration as any other feat of science or engineering, that capability translates into tangible value in an uncomplicated manner: The model generates predictive scores, which in turn drive millions of operational decisions.

For UPS, Jack used ML to predict package deliveries in order to optimize those deliveries. This kind of use case for prediction is clear-cut—and yet also momentous, even historic. It's a strategy that drives a vast range of innovation, improving practically all the main things that organizations do, all the largest-scale operations that make the world go 'round. After all, the universal key to driving better decisions is to calculate risks and likelihoods—the chances that a customer will cancel, a debtor will default, a component will fail, a transaction will turn out to be fraudulent, or a medical image conveys a positive diagnosis.

ML is the world's most important technology. This isn't only because it's so widely applicable. It's also because it offers a novel boost that can't be found elsewhere, a critical edge for what is becoming a final battleground of business: process optimization. As products and services become commoditized and organizations become increasingly homogenous in their operations, ML has come of age as a core enterprise practice necessary to sustain competitive advantage. To deploy ML is to participate in the latest evolutionary step of the Information Age.

But great gains come only by great disruption. Jack's system would improve an enormous delivery operation—but to realize this value, you'd need to make enormous changes by actually deploying it "in the field." Jack needed to sell his superiors on reshaping the entrenched procedure of more than 55,000 delivery personnel.

When Greatness Is Too Big to See Up Close

Jack's proposal also faced another dire challenge, one that's universal to all ML projects: Success only becomes apparent after tracking multitudes of cases over time. With a traditional invention, you just turn it on and see it work. You immediately witness its power right before your eyes. Press the gas pedal and the car moves. Press send and your friend receives your email message. And let's take a moment to appreciate how your windshield wiper pauses beautifully between intermittent wipes when there's only a sprinkle of rain.

But when an enterprise deploys ML, the effect is not immediately observable. It doesn't operate as a single device or take effect in a single moment. Instead, its value accumulates as it drives many decisions, such as which customer to market to, which risky debtor to lend money to, or which delivery address to plan for. You only see the benefit of changing major operations after tallying many cases over time. So, as powerful as it may be, the story is more abstract.

For many folks, this value proposition doesn't click as quickly as it would for a newly invented device. Some just aren't mentally prepared to embrace this kind of operational overhaul as the pivotal innovation that it is—one that's just as consequential as the various gizmos that have revolutionized our lives.

Besides, the decision makers who manage the world's well-oiled machines naturally resist change. They're in the business of avoiding risk rather than overcoming it. So, the inertia that an innovator feels holding them back doesn't only stem from some capricious, abominable resistance to change. The resistance is a safety measure. It's for good reason that the boss has never been prone to cultivate an interest in grasping highfalutin calculations like Jack's. Given the priorities of keeping the boat afloat, the powers that be barely even have the bandwidth to take a look.

Jack took stock of the challenge before him. Sometimes it seemed like only another fearless innovator would have the kind of vision needed to buy into this high-impact proposal. But at a long-standing Fortune

500, the head honchos just ain't that type. How do you snap them out of their corporate trance? Jack weighed his options.

One sure bet was that, if it's sexy, it sells. So how about packaging this up as the ultimate "secret sauce"?

Sexy but Vague: Artificial Intelligence

The ML industry has bitten forbidden fruit: It has chosen to promote itself as *AI*, an ill-defined umbrella term that includes ML within its malleable scope. This tends to mislead, especially when discussing a more typical, practical ML initiative designed to improve business operations and not, for example, meant to generate humanlike writing or to achieve human-level "intelligence."

While the world largely knows of ML as AI—thus the title of this book, *The AI Playbook*—the term *AI* is also how the world largely misunderstands ML. Because AI alludes to "intelligence," which is stubbornly nebulous when describing a technology, the term tends to overstate and fetishize rather than pitching the technology's concrete value. AI is sometimes used to specifically refer to ML or another kind of technology like chatbots or rule-based systems—but in many other uses, the term hints at exaggerated capabilities.

Vendors, consultants, and, chances are, some of your colleagues employ the AI brand rather than clearly advertising, without obfuscation, what an ML project actually offers. After all, plenty of folks with a budget have ears that perk up when they hear how advanced and "intelligent" a technology is, even without seeing precisely how it will improve business operations. So that route could serve to pad your wallet, at least in the short term.

But it can't last. The ML industry had better tone this down or we're all going to pay dearly. Glamorizing the core technology takes the focus off its concrete value, the specific way its deployment can improve operations. When that deployment isn't central to the plan, the plan is unlikely to come to fruition. Instead, the organization must consider the value proposition for a candidate project and buy into the project for that tangible value. Then, a very particular change management process must commence from the project's onset. Otherwise, you're prone to develop a model that never gets launched—which is the most common way ML projects fail.

Logistics Is Bringing Sexy Back

The best [ML] use cases for big business are, frankly, the most mundane ones.

—Caroline Zaborowski, astrophysicist turned data scientist

Jack is the impeccable hero of my story, so you can bet he was too prudent to garnish his pitch with flowery "AI" talk. He knew that a change affecting 16 million deliveries a day would have to be sold quite concretely. In fact, he didn't even call it *ML*, *predictive analytics*, or *data science*. Instead, he went with just about the most boring word there is for it: *operations research*.

But boring is exciting. This project was the kind of large-scale optimization that reduces tons of carbon emissions and makes tons of money. It promised large-scale change and tangible gains.

The reverse is also true: Some seemingly sexy projects have been slow to transform business. They attract a lot of attention with impressive capabilities that promise to deliver value in the long run, but so far they've enacted little to no change and they won't be moneymakers any time soon. Someday, fully autonomous cars will save countless lives, but impediments to their wide-scale deployment prevail, with some estimating that it will take decades to achieve. Likewise, IBM's computer that defeated the humans on the quiz show *Jeopardy!* excited me in 2011 like no technology ever had—but its specialized skill does not readily generalize to real-world tasks. Similarly, when ML conquers chess, Go, and complex video games, it impresses the best of us—but ML's value is captured only when it's applied practically. And most prominent of all, *generative AI* systems, which are built with ML, generate images and text—often in such an adept and seemingly humanlike manner as to give you the impression that they embody an "under-standing" of human concepts and that they can express these concepts with language and images. When employed to assist humans with cre-ative tasks, generative AI may well prove valuable to the enterprise, but to date it hasn't typically been utilized to boost enterprise efficiencies in the straightforward manner adopted by the use cases covered here.

Instead of the glitz, get excited about the measurable impact of run-ning established large-scale operations more effectively! Jack's story took place at the United Parcel Service, a sturdy complement to the US Postal Service for more than a century. We're talking about the world's largest courier, with higher revenue than even FedEx. This isn't some hot new tech company. No, this is precisely the kind of older, estab-lished firm that runs society's essential operations—entrenched proc-esses begging to be streamlined, even while many in charge fight tooth and nail against change.

Who knew that optimizing brick and mortar logistics was so sexy?

Jack's job title plainly reaffirmed this point: *senior director of process management at UPS*. He wasn't the "director of AI"—or of any technol-ogy whatsoever. His focus was on the ends—process improvements—not the means, on the business goal rather than the technical solution. Having been at the company for more than three decades, he was in charge of operations technology and oversaw six divisions. He didn't sit among the upper echelons of executives whom he now had to con-vince for approval. He worked on operations directly, right where it counts. He was situated to enact operational change personally.

But Jack's superior had given him the cold shoulder. Was this project such a great idea after all? How do you distinguish viable progress from overly radical upheaval? First, let's dive into how Jack's system worked.

Planning for Tomorrow with Incomplete Info

Imagine that you run a typical shipping center where fifty-five trucks leave every morning, each tasked with delivering 300 packages that day. Your job is to decide exactly how to distribute these 16,500 packages among the trucks so that the overall operation requires as few miles and driver hours as possible. To complicate things, some deliveries are committed for a specific time of day, plus no driver's shift can extend too long. *No pressure.*

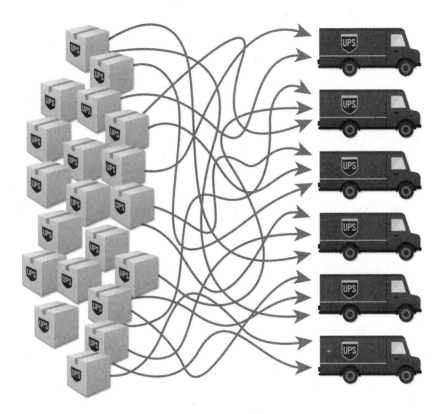

Packages are assigned to delivery trucks at a shipping center.

Now multiply this problem by 1,000. The system you develop must handle these logistics every day for 1,000 shipping centers in the United States. Across this mammoth operation, every moment counts. One minute per driver per day costs $14.5 million per year. Likewise, one mile is worth $50 million. *Really, though, no pressure.*

Overall, millions of gallons of fuel and thousands of metric tons of emissions are on the line annually. *Okay, maybe there's a little pressure.*

But here's the real kicker: The system must work with incomplete information. Shipping centers must begin the lengthy process of planning and loading the trucks before all of tomorrow's deliveries have become known. Many delivery destinations don't become apparent until the wee hours of the morning.

Jack calls this the *Delivery Paradox*. You can't optimally plan the truck-loading until you know all the deliveries that will need to be made. But by the time you know all the deliveries, you've run out of time to load the trucks.

This enormously complicates the problem. After all, every package matters for the overall plan. If an unforeseen last-minute package shows up after the trucks are loaded, it could add miles to a truck's existing plan. If you'd known earlier, you might have distributed the packages completely differently among trucks. But you're out of time. The fully loaded trucks are headed out and redistributing the packages would take too long.

Jack recognized that the Delivery Paradox was a central dilemma since shipping centers faced a plethora of unforeseen packages every day. At the time, up to 30 percent of deliveries still weren't in the system when the planning for the next day had to begin. This was because many packages that arrived on overnight flights had missing or only partial tracking information. Some shipping customers were late to upload data about their shipments or used noncompliant or glitchy systems to do so. Unexpected delays caused by factors like the weather could be slow to percolate. Throughout the long night of loading, some "dumb" packages would even show up without proper coding, so handlers then had to manually enter the destination address on the spot.

Much of this information latency persists today—it's largely unavoidable, despite various improvements UPS has made to its systems. For example, suppose that a package will fly this afternoon from the West Coast to an East-Coast shipping center for delivery tomorrow morning. If the East-Coast center begins its planning at midday, it's still too early on the West Coast for the destination address to have been uploaded. As another example, even if *all* delivery addresses have been uploaded, the number of stops each truck will need to make, each costing precious time, is often unknown until the deliveries are actually made, since, for example, a large building or strip mall with multiple recipients could turn out to require multiple stops. On top of all this, some broad-strokes planning must be completed *days* ahead, for example, to book the right number of drivers.

In a system as complex as UPS's internal package network, uncertainty is an inherent predicament. The antidote is prediction.

Predicting Tomorrow's Deliveries

Jack's system, named Package Flow Technology (PFT), predicts tomorrow's package deliveries so that it can plan for them. These predicted deliveries augment the list of known packages.

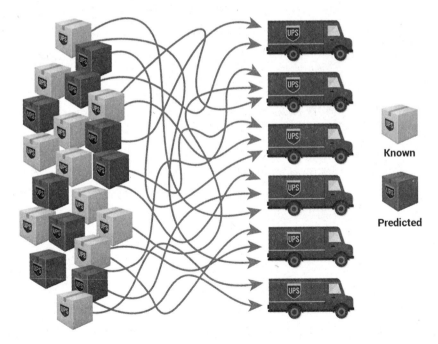

Known

Predicted

A combination of known and predicted deliveries is assigned to delivery trucks at a shipping center.

PFT can form a complete plan with this augmented batch of delivery destinations and trigger the overnight loading process with time to spare. Trucks are typically loaded from about 4:00 a.m. to 7:00 a.m., so the planning must begin earlier—in the evening or even during daytime hours for some shipping centers.

Let's look at the mechanics of how this batch of delivery predictions is formed. First, a predictive model generates each individual prediction, one at a time.

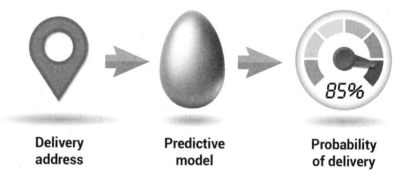

| **Delivery address** | **Predictive model** | **Probability of delivery** |

The model, shown here as a golden egg, was generated from data for this very purpose. It encodes patterns learned from the past that now serve to put odds on what will happen in the future—whether a given address will have a package coming.

The model is applied repeatedly, performing its calculations for each possible delivery address. For the United States, that's 200 million predictions. Then, all the most probable destinations—say, those scored as being more than 80 percent likely—are combined with the list of known destinations.

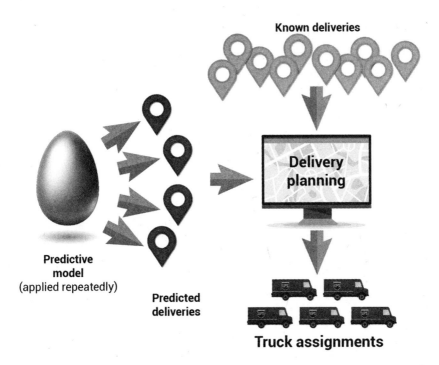

Known deliveries

Delivery planning

Predictive model
(applied repeatedly)

Predicted deliveries

Truck assignments

UPS's PFT augments the list of known packages set for delivery with a list of predicted deliveries in order to make a plan in time to load the delivery trucks.

PFT regularly updates these predictions—roughly every two minutes—until the trucks head out. Throughout the overnight loading process, some predicted deliveries become known, since the package actually shows up—plus, other unforeseen packages also come in. The system revises the plan accordingly. As a result, some packages may be moved from one truck to another, but most of the prediction-based plan remains intact without imposing time-consuming changes. By morning, any incorrectly predicted deliveries that never materialized as real packages are dropped from the plan. In the end, as the trucks head out with their packages, they no longer need delivery predictions—but the predictions are what got them well planned and fully loaded in time for that day's expedition.

Predictions Run the World

Beyond UPS, the same fundamental idea applies far and wide. The capacity to predict on a case-by-case basis is the Holy Grail because it pertains everywhere, driving decisions such as which customer to target for marketing, which patient to triage, which transaction to audit for fraud, which building to inspect for risk of fire, and which product to replenish in a supply chain.

The story is universal: Businesses and other organizations need prediction. Prediction requires ML. And ML depends on data.

Putting that in reverse, the flow runs from left to right in this sequence:

data → machine learning → model → predictions → operations

We have data, we give it to ML, it makes models that predict, and we use the predictions to drive operations more optimally.

Data fuels prediction because it embodies experience from which to learn. Looking back, an organization knows, for example, who bought what, which transactions turned out to be fraudulent, and which buildings burned. And UPS knows every shipment they've made. In making predictions with ML, a company applies what it has learned from this experience.

Let's be real: It's not a magic crystal ball. Perfect prediction is not possible—but it's also not necessary. Even lousy predictions that are at least better than guessing often deliver a tremendous systematic benefit. After all, business is a numbers game. Tipping the odds even a bit in our favor generates an enormous impact.

Going All In

In order to realize the true potential of package prediction, Jack would have to sell big change to UPS executives. By now, the PFT system had been in place for a few years, but it was only being used to divide up the packages among the trucks so that each truck had a well-designed delivery area, one that an efficient route could potentially cover.

Jack was now pitching to Chuck an additional optimization that promised even more significant gains: prescribing those efficient routes so that the drivers would follow them. Each truck was assigned a batch of packages that it *could* deliver efficiently—but that didn't necessarily mean that it *would*. The drivers needed to be told where to drive. Jack's team had shown that, for routing a truckload of deliveries, machines beat humans.

So Jack pitched this tremendous change to the executive, Chuck. He said, let's tell our 55,000 drivers to follow the computer rather than their gut. That's the only way to fully realize the efficiency promised by package prediction.

But as you know, Jack's pitch flopped.

Show, Don't Tell

Chuck didn't give a truck. He sat there, unaffected, blinking at Jack.

The next move? A stiff upper lip. Jack was deflated but not deterred. He'd already grown a thick skin from weathering plenty of "ridicule and violent opposition," as he puts it. Jack knew that selling a big idea meant navigating treacherous waters. To become data-driven, "you want people to change decisions they're making today," Jack proclaimed years later, during a keynote address at the Machine Learning Week conference. "You really have to understand change management."

Jack knew that people respond more to tangible experiences than abstractions. If they can observe an innovation in action before their eyes, then they're more likely to feel the power.

So he took Chuck for a ride. Literally. The very next day, they drove a stop-by-stop delivery route navigated by Jack's prototype. This was the first time an executive had hit the road to experience ORION—which stands for On-Road Integrated Optimization and Navigation.

A little while into the excursion, the system made a counterintuitive choice. It skipped right past three delivery addresses, leaving them for later. A human driver would never think to do this. Jack explained how this short-term sacrifice improved the overall route, ensuring that the earlier time commitments of other deliveries would be met.

Chuck put up a hand to say, "Give me a minute." Jack saw wheels turning. Then a lightbulb switched on. "This is big, isn't it?" Chuck realized with awe.

Jack had surmounted the first significant organizational hurdle toward deployment.

Precisely what kind of mental sorcery had Jack applied that persuaded this practical, risk-averse decision maker to accept disruptive innovation? After all, you can only definitively know the benefit of improving major operations after tracking performance over time—you can't see it in any one moment or even after navigating one driver's entire delivery route.

Visceral experience inspires like no PowerPoint. Numbers prove, but a journey compels. Chuck felt the magnitude of 10 million miles only after physically driving a few. By deftly navigating this human factor, Jack reached a turning point in the greenlighting of progress. Good thing he'd been a psychology major back in college.

Hit the Road, Jack

From that point onward, it still cost Jack years of tireless perseverance to get to full deployment. There were more executives to convince. And there were lengthy trial deployments at various shipping centers across the country, which slowly proved the value beyond a shadow of a doubt. As the trials became permanent rollouts, Jack developed a rigorous process to ensure each shipping center adopted the new system effectively. In the end, managing the complete national deployment required a team of more than 700 people.

Ultimately, Jack bestowed this contribution upon UPS as his swan song. He retired from the company in 2019, and yet his legacy continues to achieve astounding gains to this day. The use of ML to predict deliveries is a central component of the overall optimization system, one that, in his best estimation, contributes 10 percent of the overall savings achieved:

UPS Systems: ORION and Package Flow Technology (which work together)

Annual savings due specifically to delivery-prediction (estimated):
18.5 million miles
$35+ million

Annual savings achieved by the combined systems:
185 million miles
$350+ million
8 million gallons of fuel
185,000 metric tons of emissions

It's fair to say that Jack shot the moon. His work has received over a dozen industry awards and several high-profile TV and magazine spotlights.

A few months after retiring, Jack chatted with a UPS driver who was delivering a package to his home. For better or worse, Jack had changed every moment of this guy's workday. After all, it's the drivers at large who bear the burden of all these gains. Their time and mileage are used more efficiently, which means they wind up making more deliveries per hour all day long. Some have lodged complaints, including more tenured drivers, who may be the ones most resistant to change.

"Did you work on ORION?" the driver asked. This guy had no idea to whom he was speaking. It was a bit like asking Thomas Edison if he'd had anything to do with the lightbulb. Jack identified himself and braced for impact.

"I love ORION!" the driver exclaimed. "I'm new, only been at UPS for a year. It thinks for me. I don't have to worry about meeting delivery times. It takes the stress away."

These days, Jack reflects on the significance of his contribution. "I really am proud that, instead of a trucking company with technology," he says, "we're now a technology company with trucks."

ML Done Right and Done Wrong

Jack's story is exceptional. Beyond capturing value for UPS, he trailblazed for a new business paradigm, establishing best practices for greenlighting and executing on the massive operational overhaul that is ML deployment. Jack's work defines a blueprint for effective ML leadership.

Here's what Jack got right:

- *Value-obsessed.* Jack evangelized the concrete gains of ML deployment rather than the glitz of the technology itself.
- *Launch-focused.* He ruthlessly pursued a strategic path to deployment, instead of assuming that the value of deploying the system would be self-evident and taken on by others.
- *End-to-end ownership.* He took ownership of the entire, full-scoped process—from developing to evangelizing to testing to launching.

Years have passed, but getting these right is still not the norm. Instead, many companies flounder, neglecting to recognize the end-to-end business practice needed to deploy ML into the field.

A Tale of Two Technologies

> It was the best of times, it was the worst of times, it was the age of wisdom, it was the age of foolishness.
>
> —Charles Dickens

On the one hand, ML is "the most important general-purpose technology of our era," as *Harvard Business Review* astutely put it. On the other hand, it's the most misunderstood and mismanaged. Misfire after misfire, many ML projects go amiss in their mission to deploy, their models destined only to collect dust.

Capitalizing on this technology is critical—but it's notoriously difficult to launch. Many ML projects never progress beyond the modeling, the number-crunching phase. Industry surveys repeatedly show that most new ML initiatives don't make it to deployment, where the value would be realized.

Hype contributes to this problem. ML is mythologized, misconstrued as "intelligent" when it is not. It's also mismeasured as "highly accurate," even when that notion is irrelevant and misleading. For now, these adulations largely drown out the words of consternation, but those words are bound to increase in volume.

Take self-driving cars. In the most publicly visible cautionary tale about ML hype, overzealous promises have led to slamming on the brakes and slowing progress. As the *Guardian* put it, "The driverless car revolution has stalled." This is a shame, as the concept promises greatness. Someday, it will prove to be a revolutionary application of ML that greatly reduces traffic fatalities. This will require a lengthy "transformation that is going to happen over 30 years and possibly longer," according Chris Urmson, formerly the CTO of Google's self-driving team and now the CEO of Aurora, which bought out Uber's self-driving unit. But in the mid-2010s, the investment and fanatical hype, including grandiose tweets by Tesla CEO Elon Musk, reached a premature fever pitch. The advent of truly impressive driver *assistance* capabilities were branded as "Full Self-Driving" and advertised as being on the brink of widespread, completely autonomous driving—that is, self-driving that allows you to nap in the back seat. Expectations grew, followed by . . . a conspicuous absence of self-driving cars. Disenchantment took hold and by the early 2020s investments had dried up considerably. Self-driving is doomed to be this decade's jetpack.

What went wrong? Underplanning is an understatement. It wasn't so much a matter of overselling ML itself, that is, of exaggerating how well predictive models can, for example, identify pedestrians and stop signs. Instead, the greater problem was the dramatic downplaying of deployment complexity. Only a comprehensive, deliberate plan could possibly manage the inevitable string of impediments that arise while slowly releasing such vehicles into the world. After all, we're talking about ML models autonomously navigating large, heavy objects through the midst of our crowded cities! One tech journalist poignantly dubbed them "self-driving bullets." When it comes to operationalizing ML, autonomous driving is *literally* where the rubber hits the road. More than any other ML initiative, it demands a shrewd, incremental deployment plan that doesn't promise unrealistic timelines.

How to Get It Right

It's the same disappointing story with many ML projects, even though you don't usually face a deployment challenge nearly as great as when installing models into autonomous vehicles. Likewise, you also don't usually face the complexity of optimizing UPS's operations. Most ML projects have it much easier. And yet greenlighting and managing model integration still turns out to be much harder than expected.

In this book, I present a strategic and tactical playbook for launching ML, a six-step business discipline to run an ML project so that it successfully deploys. I call this practice *bizML*. It overcomes ML's common failure to launch, surmounting the hurdles to deployment by planning for it from the get-go, in detail, even before the hands-on data work begins.

Along the way, I also cover the semi-technical background knowledge everyone participating in an ML project needs—in a friendly, accessible way that anyone can understand. With everyone on the same page, a multidisciplinary team can collaborate deeply throughout the entire end-to-end project.

The next chapter, chapter 0, overviews the bizML paradigm, kicking off with a cautionary tale from my own consulting practice. Then, chapters 1 through 6 cover the six-step discipline. Last, the conclusion covers some final ingredients needed for planning an ML project, including the project's staffing, timeline, ongoing upkeep, and ethical considerations.

0   BizML

Six Steps to Machine Learning Deployment

> Why start with a "chapter 0"? This chapter introduces *bizML*, the six-step playbook covered by chapters 1 through 6. BizML maps out the strategic practice needed to get machine learning launched—that is, not only to perform number crunching on data, but also to operationally deploy the results. To understand why this approach is needed, this chapter addresses several pressing questions: Why do most ML projects fail to deploy? Why is it so important to plan ML projects backward? Why must business leaders possess semi-technical know-how, even if they're not data scientists? And who should lead the project in the first place?

It's a crisis. Humanity's latest, greatest invention is stalling right out of the gate. Machine learning projects routinely fail.

This cannot stand. The world needs ML. It combats our most significant risks—including wildfires, climate change, pandemics, and child abuse. It boosts sales, cuts costs, prevents fraud, streamlines manufacturing, and strengthens healthcare.

But most ML initiatives die an early death: They stall before deploying. These misfires cost us dearly. Who's to blame?

Blame data scientists like me.

My regrettable behavior began years ago. I was visiting the hip San Francisco offices of gay.com—now defunct, but then the most popular gay dating site in the United States. This was the first significant client engagement I'd landed as a newly minted independent consultant.

"Sorry for making you wait," the VP said to me, scurrying out of a conference room. She showed me a guilty grin—but the expression on her face wasn't about time; it was about money. "We have millions of dollars sitting idle in a checking account . . . Oops! So, we had to decide where to, like, invest it or something."

Flush with cash, she signed my contract renewal at three times the rate that I'd even hoped to receive as a new consultant. "This is the last one," she warned, as if issuing a Twinkie to a ten-year-old.

I could have jumped for joy. Children need to play, surgeons need to cut, and data scientists need to model. I'd fallen head over heels for ML more than a decade earlier, but I'd so far pursued my passion only through academic research and teaching. Now I had a real company paying me real money to prove ML's value to the real world.

The Potential of Prediction

Like most techies, I agreed with something Jack Levis had once said: "The business drives technology; the technology doesn't drive the business." But, for an excited data scientist, it can also work the other way around. My project pursued a worthy business goal, but it served me as well. I got to flex my modeling muscles in the name of that pursuit.

Gay.com's business goal was one of the most rudimentary of all: Retain more customers. After all, a customer saved is a customer earned. It's a well-known marketing rule of thumb that holding on to an existing customer is far more cost effective than acquiring a new one "off the street."

Prediction makes customer retention possible. If gay.com targeted customers likely to leave with marketing contact and that managed to turn around just 5 percent of those who would otherwise cancel their paid membership, the company would annually gain an estimated $862,000 in *customer lifetime value*, the additional revenue to come from those saved customers.

Why such great potential? This dating site was for brief flings. On other sites populated with relationship-seekers, users were prone to stick it out (until eventually finding a partner and canceling, after which there was little chance of winning them back). With gay.com, up to 80 percent of paying members were canceling before their next auto-renewal. I saw that behavior in the numbers, but at first I was clueless about why. As I crunched data behind the scenes, I was removed from the website's look and feel. I knew it catered to male users, but I'd only catch a glance at the front end every once in a while. It was plastered with hot dudes. At one point, someone spelled it out for me: "This is where guys go for casual hookups."

As such, business was booming. What gay.com lacked in customer longevity it made up for in popularity and a constant influx of new subscribers. This left a healthy base of 145,000 subscribers paying for a premium membership, plus several times as many free accounts.

This high turnover rate presented both an opportunity and a challenge. Given the torrent of customers flowing in and out, all we had to do was retain some small portion. Just dipping a small bucket into that wide river would be a great win. But changing the mind of a customer who'd otherwise defect is expensive, usually achieved by offering them a tempting discount. A company can't afford to offer a discount to its entire customer base.

Prediction is the only recourse. A company focuses its outreach, offering a retention discount only to the customers most likely to cancel. This common use of ML, called *churn modeling*, pays off for many organizations. It's popular, for example, with cell phone carriers. Telenor, the world's seventh largest—with over 150 million subscribers—deployed churn models to boost the return on its retention efforts by a factor of eleven.

So, all I had to do was develop the predictive model and explain its potential to gay.com. Surely they'd use it to drive a marketing campaign and capture this unrealized profit.

The Two Main Technical Steps of Machine Learning

Those two steps—developing a model and then using it—are universal. They're the two main technical steps for any ML project. First, a *modeling method,* a.k.a. *ML algorithm,* takes data as input and processes it to generate a predictive model.

Data   **Machine   **Predictive
                             learning**   model**

Machine learning generates a predictive model from data.

The model, depicted here as a golden egg, is the thing that's been "learned" from data. For gay.com, I generated *decision tree* models, which are made up of if-then rules, such as:

```
IF the thing the customer was trying to do when they
   got upgraded from a free to a paid membership
   level was trying to chat with another user
AND
The customer signed up for a paid membership fewer
   than 238 days ago
AND
They last had a failed login attempt within the last
   2 days
THEN the probability of cancelation is 43%.
```

That's the fun part. When machines automatically discover historical patterns—either as if-then rules or as more sophisticated mathematical formulas—you're witnessing the most exciting, fascinating, and far-reaching kind of technology: software that learns.

Once created—or "learned"—the model's purpose is to generate predictive scores (probabilities) for current customers who haven't yet churned on a case-by-case basis. So, for the second step, we use it to do so, thereby applying what's been learned. That's called *scoring*. Scoring with a model and then acting on the score is called model *deployment*.

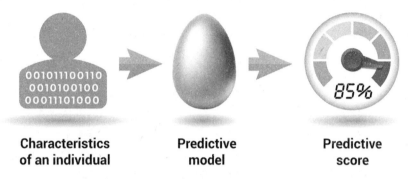

| **Characteristics of an individual** | **Predictive model** | **Predictive score** |

Scoring: A model generates a prediction for an individual.

This is the launch, the main event, the whole point. The first step may have been the "rocket science" part, but eventually the time must come to launch the rocket. In the deployment step, we dispatch the model into the field where it generates predictions that drive operations—such as marketing outreach to retain customers. Model deployment is also referred to as *operationalization, integration, implementation,* or "putting a model into production."

For example, once the model's been deployed, when the rule outlined above "fires" for a customer and therefore scores the customer with a 43 percent probability of cancellation—which is relatively high—it triggers an email that says, "We love you as a customer, here's a discount code."

That's what I pitched to gay.com. After developing the model, I knew I had to make the business case, so I worked up a nice Power-Point presentation to show the potential of my model and its estimated returns—if only they'd use it.

Machine Learning Failure Is Usually Human Failure

> Individual use of technology does not always translate into knowing how to use these tools effectively within or as an organization.
>
> —Gerald Kane et al., *The Technology Fallacy: How People Are the Real Key to Digital Transformation*

> Statisticians, like artists, have the bad habit of falling in love with their models.
>
> —Famed statistics professor George Box

But gay.com didn't deploy my model. Instead, they complimented it and placed it on the back burner. The VP assured me that my proposal was interesting and that they would take it into consideration. She thanked me for my consulting engagement, which had drawn to a close. No more Twinkies for me—and no value achieved for them.

I quickly shrugged it off. As a new consultant in a field that was just beginning to catch fire, I could only afford a moment to feel perplexed and disappointed. I'd done my part and it was their loss.

But today, after two more decades of consulting, the lesson has sunk in. Unfortunately, this lesson is still just as relevant and hard-earned today, perhaps in part because it's a paradoxical one:

> **The ML Paradox**: For this advanced technology to succeed, we now need improvements in humans—in the way of understanding and leadership—more than in the technology itself.

The ML Paradox is a special case of what James Bessen calls the *Automation Paradox*. "When computers start doing the work of people," he wrote in the *Atlantic*, "the need for people often increases."

It takes a holistic view—one that integrates business- and technology-side perspectives—to sell, educate on, socialize, and lead ML projects. Lacking this, organizations often fail to bridge the business/tech "culture gap." On the one hand, data scientists, who perform the model development step, compromise the value of their work by fixating solely on data science. As a rule, they prefer to be left to their highly technical area of expertise, not bothered with "mundane" managerial activities. There's a tendency for the data scientist to take the deployment of their model for granted. Its value is self-evident—how could it not be put to real use? With that mindset, they enthusiastically jump past a rigorous business process and straight into the modeling. In most cases, the resulting model only collects dust.

On the other hand, many business professionals—especially those already inclined to forgo the particulars as "too technical"—have been seduced into seeing this stunning technology as a panacea that solves problems on its own. To them, there's no need to get into the details, since the tech is intrinsically valuable and the details belong only within the purview of data scientists. Ultimately, when faced with the operational change that a deployed model would incur, it's a tough sell. Taken off-guard, the stakeholder hesitates before altering the very way in which the company maintains its profitability.

With no one taking proactive ownership, the hose and the faucet fail to connect. The irony is undeniable: All parties tend to focus more on the technology itself than how it should deploy. *This is like being more excited about the development of a rocket than its launch.*

Many Models Never Deploy: An Industry-Wide Problem

At companies where there is no framework for the operationalization of models, PowerPoint is where models go to die!

—Hulya Farinas, Director of Data Science, Fitbit

How many models fail to deploy? A majority of data scientists say that, in their work, it's between 80 and 100 percent. Across their projects, only 0–20 percent of models deploy. Thankfully, the remaining data scientists experience higher success rates. These observations come from a survey I conducted with KDnuggets, a seminal analytics news site popular with data scientists, as summarized by this bar graph.

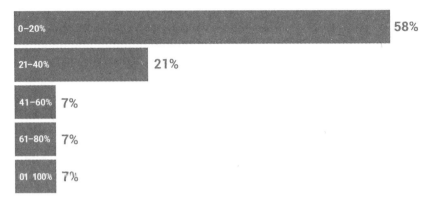

0–20%	58%
21–40%	21%
41–60%	7%
61–80%	7%
01 100%	7%

Survey responses to the question, "What percentage of ML models (created by you or your colleagues with the intention of being deployed) have actually been deployed?" Total respondents: 114.

Other industry research aligns with this dismal result. The industry-leading Data Science Survey run by ML consultancy Rexer Analytics showed that only 11 percent of data scientists say their models always deploy. Managers follow suit, indicating that "only 10% of companies obtain significant financial benefits from AI technologies," according to research from MIT Sloan Management and the Boston Consulting Group. Likewise, an analyst at the research firm Gartner estimated that close to 85 percent of big data projects fail.

Still, ML is by no means a flop. If 15–25 percent of the world's many projects deploy, that's far from nothing. Predictive models positively impact our lives on a daily basis, delivering more relevant content—for example, by empowering spam filters and Google search results—drastically reducing credit card fraud, and much more. The Machine Learning Week conferences I've been running since 2009 are built on loads of positive case studies from Fortune 500s and beyond.

In fact, for many ML projects, success is relatively likely. For example, at Big Tech firms, experienced staff and the sheer power of abundant resources often align the stars for deployment. The same is true if you're developing a newly updated model at a bank that's already been deploying credit risk models for years. Moreover, rare innovators like Jack Levis at UPS succeed by leading unusually deployment-oriented projects.

But if you're pioneering a deployment that's new to your organization and luck hasn't aligned the stars, you may have your work cut out for you. McKinsey's AI Index reveals that the "divide between AI leaders and the majority of companies still struggling to capitalize on the technology" is only widening.

A rising tide of unfavorable buzz and anecdotes bemoans this disappointing truth. When I posted our survey results, experts concurred. For example, Analytics leader Armin Kakas, who's managed analytics at GE Capital, Molson Coors, Best Buy, and American Tire Distributors, chimed in to say, "Over the years, I've led or had oversight of many enterprise-level analytics initiatives. . . . If I'm generous, I'd say about one in five succeeded and had some level of value realization for the company."

Consultants fight the same battle. Famed *Digital Decisioning* author James Taylor—one of the few consultants I know who run analytics initiatives from a business vantage rather than a data science one—has seen the same thing time and again. "In conversations I've had with many companies," he says, "when I start asking about machine learning deployment, with business results not technical ones, fully rolled out not just pilots, there are not a lot of good examples. I say that my definition of project success is when a model has been developed and deployed such that it has *created*—note the past tense—business value for the organization that paid for it. When you impose that criterion, man, it's quiet out there."

The power is stuck in a PowerPoint. We're slow to put it to good use. Then-MIT Sloan professor Erik Brynjolfsson put it plainly in a TED Talk: "Technology alone is not enough. Technology is not destiny. We shape our destiny, and just as the earlier generations of managers needed to redesign their factories, we're going to need to reinvent our organizations. . . . We're not doing as well at that job as we should be."

ML's broader success isn't delayed pending some new technical breakthrough. It's on hold until the tech breaks through—until it achieves organizational acceptance and operational adoption. If you're embarking upon a new ML initiative and you don't go above and beyond to undertake a very particular organizational practice to run the project, the risk is high that your project will slip into the same pattern that has so often led to failure for others.

Before I outline that practice, let's look at why so many ML projects struggle to deploy.

They *Can't* Deploy . . . or Just *Won't*?

The ML industry has nailed the *development* of potentially valuable models, but not their *deployment*. A report prepared by the *AI Journal* based on surveys by Sapio Research showed that the top pain point for data teams is "Delivering business impact now through AI." Ninety-six percent of those surveyed checked that box. That challenge beat out a long list of broader data issues outside the scope of AI per se, including data security, regulatory compliance, and various technical and infrastructure challenges.

But when presented with a model, business leaders refuse to deploy. They just say no. The disappointed data scientist is left wondering, "You can't . . . or you won't?"

It's a mixture of both, according to another question asked by my survey with KDnuggets, as summarized by this bar graph.

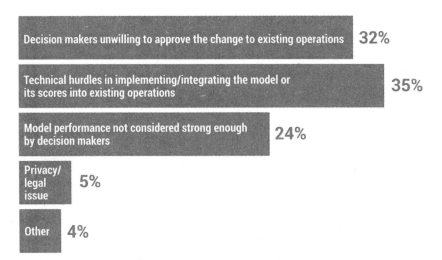

Survey responses to the question, "What is the main impediment to model deployment?" Total respondents: 114.

Technical hurdles mean that they *can't*. A lack of approval—including when decision makers don't consider model performance strong enough or when there are privacy or legal issues—means that they *won't*.

Another survey also told this "some can't and some won't" story. After ML consultancy Rexer Analytics' survey of data scientists asked why models intended for deployment don't get there, founder Karl Rexer told me that respondents wrote in two main reasons: "The organization lacks the proper infrastructure needed for deployment" and "People in the organization don't understand the value of ML." Unsurprisingly, the latter group of data scientists—the "won'ts" rather than the "can'ts"—sound the most frustrated, Karl says.

Whether they can't or they won't, the lack of a well-established business practice is almost always to blame. Technical challenges abound for deployment, but they don't stand in the way so long as project leaders anticipate and plan for them. With a plan that provides the time and resources needed to handle model implementation—sometimes, major construction—deployment will proceed. Ultimately, it's not so much that they can't but that they won't.

We turn now to the remedy, the proactive planning process every ML project needs so that they can and they will.

The Solution: BizML

Most analytics and AI projects fail because operationalization is only addressed as an afterthought.

—Gartner

As we saw with UPS's story in the introduction, Jack Levis succeeded in his diehard drive to deploy by taking end-to-end ownership—from project inception to launch. That's a wide span to cover.

The same is required for any ML initiative. To run the project so it successfully launches, you must follow a business practice built on three fundamentals guidelines—the third of which relates to end-to-end ownership:

Ramp up on a semi-technical—yet accessible—side of ML. Before their involvement in an ML project, all professionals must gain some rare background knowledge, familiarity with a collection of accessible ML fundamentals. Surprisingly, some of them are seldom known even by data scientists. Chapters 1 through 6 of this book cover these fundamentals while stepping through the ML business practice.

Mandate deep collaboration between business professionals and data scientists. These two very different "species" must team up to execute each project step in order to achieve deployment. After all, deployment means radical change to existing operations. You can't assume that the decision makers will buy in easily. To keep things on track and grease the wheels for operationalization, business-side stakeholders must be enlisted to deeply collaborate with data scientists and weigh in at each project step, end to end. This includes defining performance goals, preparing the data, and developing and deploying the predictive model.

Plan ML projects backward. In the first step of that collaboration, before modeling begins, start with the end goal: precisely how ML will be deployed to improve operations. Stakeholders must approve the way in which the probabilities calculated by a model will change business processes in order to improve them. Only by declaring this up front does an ML initiative stand a chance of achieving successful deployment. This "backward" strategy is a simple and yet surprisingly underutilized trick of the trade.

To embody these three guidelines, a *knowledgeable* team must *collaboratively* follow an end-to-end practice that begins by *backward planning* for the end. I call this practice *bizML* and it consists of the following steps.

The six steps of bizML:

1. *Value: Establish the deployment goal.* This step defines the business value proposition: how ML will affect operations in order to improve them by way of the final step, model deployment.

2. *Target: Establish the prediction goal.* This step defines exactly what the model will predict for each individual case. Each detail of this matters from a business perspective.

3. *Performance: Establish the evaluation metrics.* This step defines which measures matter the most and what performance level must be achieved—how well the model must predict—for project success.

4. *Fuel: Prepare the data.* This step defines what the data must look like and gets it into that form.

5. *Algorithm: Train the model.* This step generates a predictive model from the data. The model is the thing that's "learned."

6. *Launch: Deploy the model.* This step uses the model to render predictions (probabilities)—thereby applying what's been learned to new cases—and then acts on those predictions to improve business operations.

These steps define a business practice that forges and navigates a shrewd path to ML deployment. Anyone who wishes to participate in ML projects must be familiar, no matter whether you'll play a role on the business side or the technical side. Rip out the one-page cheat sheet toward the end of this book and hang it on the wall above your desk.

If you've seen this book's table of contents, these six steps will look familiar. They're the six subsequent chapters of this book. For each step, a full chapter is dedicated to exploring the step's considerations, decision points, and challenges.

After culminating with step 6, deployment, you have finished . . . starting something new. BizML only begins an ongoing journey, a new phase of running improved operations—and of keeping things working. Once launched, a model requires upkeep: monitoring it, maintaining it, and periodically refreshing it. This book's conclusion introduces that ongoing effort.

Why the Industry Is Converging on These Six Steps

Following these six steps in this order is almost a logical inevitability. To understand why, let's start with the end. The final two culminating steps, steps 5 and 6, are the two main steps of ML, model training and deployment. BizML ushers the project through to their completion.

The step just before those two—step 4: Prepare the data—is a known requirement that always precedes model training. You must provide ML software with data in the right form in order for it to work. That step has always been an integral part of modeling projects, ever since linear regression was first applied by businesses in the 1960s.

Before the technical magic, you must perform business magic. That's where the first three steps come in. They robustly backward-plan the project—before diving into the hands-on work of prepping data, developing a model, and using it. They establish a greatly needed "pre-production" phase of pitching, socializing, and collaborating in order to jointly agree on how ML will be applied and how its performance will be evaluated. Importantly, these first steps go much further than only agreeing on a project's business objective. Business professionals, prepare to dive more deeply into the mechanics and arithmetic than you might have expected. Likewise, data scientists, prepare to reach beyond your usual sphere of techies and work closely with business-side personnel.

I have designed bizML to satisfy a dire, unmet need: The industry has not yet established a standardized practice that's well known to leaders, managers, and other business professionals who wish to apply ML. One standard established almost thirty years ago gained some traction, although almost entirely among data scientists (back then, *data miners*). Designed in 1996, it's called *CRISP-DM*: the CRoss Industry Standard Process for Data Mining. This foundational effort paved the way for standardization by laying out many of the fundamentals. However, it never gained much traction among business professionals, in part because it largely spoke the language of tech and perhaps also in part because it eventually came under the control of the ML software vendor SPSS (later acquired by IBM), limiting any potential for it to be championed by a vendor-neutral initiative.

This book's introduction of bizML represents a renewed effort to establish an updated, industry-standard playbook for running successful

ML projects that is pertinent and compelling to both business professionals and data professionals.

BizML complements CRISP-DM, so the two are compatible. CRISP-DM applies for a broader range of projects—data science projects in general—whereas bizML focuses on ML projects specifically, so this book's steps delve more deeply into predictive model-specific topics, such as a project's prediction goal, predictive performance metrics for models, and how models are deployed.

On a more contemporary front, bizML also fits in neatly with the recently christened practice *data product management*, which advocates for a "product" orientation on analytics projects that closely parallels software product management. This means developing capabilities well tuned to the needs of the customer—who, in the case of ML projects, is the person who consumes model scores, either to guide their team's decisions or to improve their operational system. By planning backward, bizML abides, putting the customer first and serving up a viable product: predictive capabilities that meet the customer's needs.

But bizML is more specialized. The wisdom of data product management pertains to all kinds of analytics projects in general. It borrows prudently from software product management's best practices to run a product's development, maintenance, and customer support. Meanwhile, bizML is designed for ML projects in particular, with all six steps specifically addressing what it takes to successfully deploy a predictive model.

Before jumping into step 1 of bizML with the next chapter, this chapter still needs to finish setting the stage. I'll now cover:

- Why ML lingo needs a new term: *bizML*
- Why the technology itself gets more hype than its launch
- How the ML industry must reframe itself
- Why *data literacy* is for everyone—like driver's education, not auto-mechanic school
- Who should lead ML projects
- What I should have done differently with the gay.com project

BizML: A Core Requirement That Had Gone Unnamed

Following all six of the steps of the bizML practice is uncommon, but hardly unheard of. Many ML projects succeed wildly, even if they're in the minority. While a well-known, established framework has been a long time coming, the ideas at the heart of the bizML framework are not new to many experienced data scientists.

And yet the folks who need it the most—business leaders and other business stakeholders—are least likely to be familiar with it. In fact, the business world in general has yet to become aware of even the need for a specialized business practice in the first place. This is understandable, since the common narrative leads them astray. AI is often oversold as an impenetrable yet exciting cure-all. Meanwhile, many data scientists far prefer to crunch numbers than to take pains to elucidate. All along, there's been no popular business book and no commonplace business school curriculum that teaches a detailed playbook for ML projects.

Most unhelpfully of all, there's not even been a name for it. No widely recognized lingo to spread the word and create a trend. Instead, the ML buzzwords that have so far gained traction pertain to technical methods, not to the business-side discipline. For one, the trending field *MLOps*—which deals with ML *operationalization*, another term for deployment—solves technical hurdles, not organizational ones. MLOps refers to an important collection of engineering "tricks of the trade" for managing and maintaining models. This should not be confused with a practice for managing humans. Although both relate to the operationalization of models, MLOps addresses the technical execution of ML projects and bizML addresses the organizational execution. A project following bizML may well employ MLOps as an invaluable approach to make sure that the company *can* deploy a model technically, but MLOps doesn't holistically address whether its leaders *will* deploy it. No technical solution alone can address the business-side challenges faced by ML projects. Instead, an effective business practice like bizML must be the dog that wags the MLOps tail.

AutoML is another popular term, but it also names technical solutions rather than organizational ones. AutoML refers to an invaluable collection of methods that automate some of the tasks traditionally performed manually by data scientists, including certain aspects of data preparation and of selecting the best modeling method (parts of steps 4 and 5, respectively).

Hence the name *bizML* for the six-step practice for running an ML project that is presented by this book. Rather than a technical practice, this is a business practice that involves technical steps.

The Origin and Cost of Hype

I predict we will see the third AI Winter within the next five years. . . . When I graduated with my PhD in AI and ML in '91, "AI" was literally a bad word. No company would consider hiring somebody who was in AI.

—Usama Fayyad, speaking at Machine Learning Week, June 2022

Wait a minute! By ordering the chapters according to the six project steps, I have written an ML book that doesn't dive into the ML itself until near the end of the book.

In fact, that's perfect. A project intended to launch ML must concern itself first with how a model will deploy and only second with the core number crunching that will generate the model—no matter how exciting and impressive that crunching may be from a scientific viewpoint.

But the data scientist fetishes core ML methods. She was born that way. Her impulse is to go "hands on" with modeling as soon as possible. Even fledgling data scientists follow suit, beginning almost invariably with hands-on courses and books that presume the training data is already prepared. And the vast majority of instructors and authors egg them on: The first step is to load data into the modeling software. This supports a false narrative that condones skipping right past the earlier steps of a project. As a result, jumping straight into the core ML itself before establishing a path to operational deployment is the most common mistake that derails ML projects.

Some straightforward economic factors further amplify this disproportionate focus on the core technology. In the analytics industry, the best way to make a killing is by selling software. That's where the margins, acquisitions, and IPOs are. But the vendors selling ML software tools aren't inclined to advertise that their products do not themselves perform operational change. They may be slow to explain that ML software takes on only limited—albeit central—technical portions of an end-to-end ML project. These vendors are incentivized to keep the focus on their technical products rather than the enterprise process.

This overfixation is going to cost us. Like it or not, we're strapped in for the ups and downs famously depicted by the Gartner hype cycle, which illustrates the expected trajectory for each new technology, from inception to maturity.

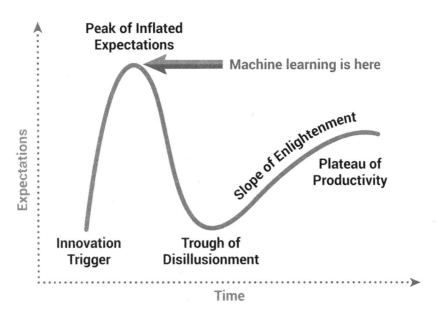

The Gartner hype cycle for technology.

At first, expectations rise as a new technology gains traction. But the hype typically goes too far, reaching the "Peak of Inflated Expectations."

With ML, we're right at that precipice, ready to tumble down to the "Trough of Disillusionment." Gartner suggests that ML has already slid partway down toward the trough, but I would argue that it still rests at the peak. Industry excitement about ML does not yet appear to have been affected by its struggling deployment rate.

ML's plunge could be worse than most. The disconnect between hype and reality seems to only be increasing, with so much more attention being given to the tech than to its deployed value. When the executives catch on, there'll be hell to pay. If we continue on the current course, the fall from grace will be steeper and the disillusionment deeper. It'll be central to a third *AI Winter*, an era of diminished excitement and funding. Our subsequent climb back up the "Slope of Enlightenment" will likely be more gradual, with productivity a longer way off.

Reframing ML

> We don't focus on the technology. To tell you the truth, I don't have a single analytics project. I've got business projects and analytics may be part of that.
> —Jack Levis

If we reframe ML, it could avoid such a dire plunge. Although I believe AI Winters are inevitable, that painful nosedive need not take ML along for the ride. By realistically and concretely communicating what ML offers—and, ideally, by calling it "ML" rather than "AI"—we can differentiate it from the often-misinformed hype that defines the AI brand and save ML from being a victim of its own hotness.

We need only reorient the focus. Don't propose an "ML project." Instead, pitch and lead a project that will improve operations, with no more than a side note mentioning ML as part of the solution. Usually, ML projects are framed in this way:

AI will improve operations.

In addition to dispensing with the term "AI," which usually compromises clarity, we must reframe the project in this way:

We will improve operations (using ML).

By way of illustration, let's apply this reframing to the stories we've covered so far—put in the briefest of terms:

We will run a new marketing process to retain customers (using ML to target those most at risk of canceling).

We will improve the efficiency of UPS's package deliveries (using ML to predict delivery destinations).

As a couple more examples, let's properly frame a *credit scoring* and a *fraud detection* project:

We will improve our bank's credit application processing (using ML to predict which applicants are most likely to default on their loan).

We will increase our detection of fraudulent transactions without increasing the auditors' workload (using ML to predict which transactions are most likely to be fraudulent).

Reframing ML projects in this way puts the business objective first, rather than the technology—and, likewise, it shifts the agency from the technology to the business. The first word of the sentence is "we" not "AI," humans not machines.

Moreover, this adjustment puts *change management* squarely on the agenda. Change management is a well-established discipline designed to facilitate operational shifts—but it can only do so if it's employed. Many ML projects don't recognize that the notion of change management applies, but model deployment means changing the very way the business operates and that change must be proactively managed like any other. By viewing the endeavor as an *enterprise* project rather than a *technology* project, folks will recognize an often-overlooked truth: *ML deployment presents a challenge that only a change management process can meet.*

ML Vendors Help, but Enterprises Lead the Industry

ML software creates *potential* value, but only a broader enterprise initiative can capture that value. An ML project succeeds not only by using an

ML software product, but by implementing a paradigm. For the business project, ML software plays a vital role, but it's only a supporting one.

With some kinds of technology, the product alone delivers value. Faster computing hardware, larger storage solutions, and streamlined database software provide value in and of themselves. But ML software is different: To improve business operations, it must be used as only one part of a broader organizational process.

After all, no software product alone could solve the kind of monumental problems that ML projects solve: *operational inefficiencies*. An ML project is a consulting gig, not a technology install. Those running the project are delivering services more than software.

Accordingly, vendors don't lead the ML industry—users do. In this way, the ML industry differs from the cellphone industry, the laptop industry, or even many forms of enterprise software, where companies compete for a pole position that defines the market by way of the products they sell. Instead, the ML industry is a movement led by the innovators within the companies that use ML. It's more like the restaurant industry, where stove manufacturers are critical but restaurateurs and chefs lead the industry.

The myth of a do-it-all *citizen data scientist* is spread in part by certain ML vendors. They do so when they sell the false narrative that their ML software product itself solves an enterprise problem. *Au contraire*— untrained users can't, unassisted, develop predictive models for new ML initiatives. ML software requires data science expertise to use and thereby create value. And capturing that value takes more than only using the product—it requires a holistic, collaborative enterprise practice. Inexperienced enterprises turn to ML vendors for guidance, but vendors lack an incentive to elucidate on the full context within which their product must be used. They often exploit that inexperience, selling a product to a customer who hasn't realized what it will take to capture value with it.

Now, if you were hoping to be wowed by technology that just "plugs in" and generates value, it's time to manage expectations. That's silver-bullet thinking. From that vantage, there's bad news:

An ML project is a business endeavor, not simply a technical one that can be handed off to data scientists to take on alone.

After all, a model is going to directly change business operations, so the project requires a wholly collaborative process driven by business needs. That's entirely unlike other data-intensive initiatives, such as deploying a data warehouse or certain business intelligence reporting solutions, which can be handed off to the IT department and revisited later to receive the results.

But the bad news is also good news:

An ML project is a business endeavor, so those involved have the opportunity to guide the process and ensure that the resulting model is actionable within the company's operational framework and has the greatest impact within the company's business model.

The centerpiece of an ML project is model deployment and the operational improvement achieved by doing so—not the use of the software that generates the model in the first place.

The Semi-Technical Background Knowledge You Need

Reframing ML will help correct the common misconception that business professionals need not become acquainted with any of its particulars. Many mentally tuck ML into a black box that only data scientists penetrate. Analytics vendors love this box, since that which is mysterious seems powerful.

But most data professionals agree that business stakeholders need to ramp up. According to the "State of Data Science" survey by the analytics vendor Anaconda, "Only 36% of people said their organization's decision-makers are very data literate and understand the stories told by visualizations and models." Despite this, only 1.6 percent of respondents name *data literacy* as their "primary area of data investment," according to the NewVantage Partners 2023 survey of senior data and analytics executives.

So, like any tech author worth their salt, I'm here to demystify. It's time to blow the lid off and become familiar with some of the inner workings.

Many business professionals balk at this suggestion. "I don't need to understand the inner workings of an engine to drive a car. I delegate all that technical stuff to the experts. Tinkering under the hood is someone else's responsibility."

Fair point—but here's how that analogy actually applies: This is driver's education, not auto-mechanic school. In order to drive, you *do* need extensive know-how, familiarity with core fundamentals such as navigation, acceleration, momentum, friction, and collisions. You must become intimately acquainted with how a car interacts with the world and how you control it.

The same goes for ML: To drive business with it, you must fully grasp its fundamentals, even if you aren't working "under the hood." In addition to learning about the six bizML steps, business professionals must gain a particular kind of *data literacy* by ramping up on certain semi-technical particulars, precisely so that they can actively weigh in on them. They include:

> **Deployment**: Precisely what's predicted and exactly how those predictions will change operations in order to improve them.
>
> **Performance**: The particular arithmetic to measure and report on how well it works—how effectively it predicts and the bottom-line business impact of using it.
>
> **Data**: How to source and prepare this "raw material"—what it needs to look like so that ML software can make use of it.
>
> **Models**: What they function to do—what they take as input and what they produce as output—plus the gist of what happens in between. As with the basics of internal combustion, the principles behind these "prediction machines" are within every professional's capacity to understand—and within every professional's purview as well. After all, they're poised to actively alter the very operations of your business.

This book covers these semi-technical fundamentals, addressing them along the way while proceeding through the six steps of bizML across the next six chapters. If you're a business professional and until now have felt these kinds of details fall outside your job description, it's time to embrace the paradox: After just a bit of ramp-up, more than anything, *non-data scientists like you are exactly what ML projects need.* With this background knowledge, you can meet data scientists halfway. They're great at creating value; now it's up to you to capture it.

Who's in Charge?

This upskilling will make you not only a valuable participant, but also the potential leader of ML projects—sometimes called the *data product manager.* BizML and the associated background knowledge empower business professionals to take the lead by providing an understanding of precisely how deployment will change operations—after all, one must know the change in order to manage it. Likewise, these learnings also empower data professionals by providing a holistic framework for aligning their technical work with business goals.

The leader doesn't have to be you, but think twice before ruling yourself out. If you've taken all this to heart, you're good to go. You recognize the value of the six-step practice and the additional background knowledge that may be new to you. The age-old proverb applies here more than anywhere: If you want something done right, you've got to do it yourself.

But if leadership doesn't interest you, there's another way you can still be the hero of your ML story: Make sure that whoever takes the lead is willing to spearhead the end-to-end business practice. If you're not proactive in this way, you run the risk of defaulting to authority structures that have so often proved themselves insufficient. While it's understandable that many business professionals defer to data scientists to run the show, that technical expertise doesn't necessarily come with the multifaceted enterprise leadership know-how that ML projects need to achieve deployment.

Meanwhile, data scientists often submit to the authority of money. When gay.com paid me to model customer cancellations, it felt like validation—but it wasn't. Just because those in charge are paying you doesn't mean they're signing up for big operational change. I've seen many of my colleagues also enjoy this kind of "prestige position," where their employer or client feels comforted and proud, knowing they're pursuing the latest, greatest technology—yet, all the while, deployment is only a distant dream.

But with a bit of upskilling, you or your colleagues can become that rare bird who, by bridging the all-too-common skills gap, is prepared to lead ML projects. If you're a business professional, you may still need to gain the semi-technical background knowledge we've discussed. And if you're a data professional, you also may have some learning to do. "Soft and business-related skills were the most significant gaps between what universities teach [data scientists] and what organizations need," according to the "State of Data Science" survey I mentioned earlier.

As the industry warms up to this upskilling process, there's no established standard for who should take on the leadership role. Instead, there's flexibility. In principle, a line-of-business manager may be the most natural fit, since the person running the operations that will be improved with ML should be the one in charge of optimizing them. This person usually owns the business objective, such as the reduction of customer churn. On the other hand, it's data scientists who often have the clearest vision of how their handiwork could generate value. For many projects, a data scientist has been the main champion, having energized the project's inception in the first place. Alternatively, an experienced leader might come from your analytics center of excellence, or your chief data officer could lead ML projects, although they're generally too busy with executive responsibilities—it's an in-depth, demanding process, as this book intends to demonstrate.

When it comes to running the project, it's what you do, not who you are. It's the process that matters, not the job title. BizML's six steps are universal—they always work, regardless of your org chart. This is a very good thing, considering that internal structure varies like mad across companies. It would be a fool's errand to prescribe any one-size-fits-all organizational structure for analytics projects. For all members of an ML project's team—including the leader—it's the job duties that count.

Whoever takes charge, they've got to facilitate deep biz/tech collaboration. In addition to managing the technical tasks, the leader must rally key stakeholders, executives, and decision makers. Only by achieving a certain critical mass of enterprise-wide engagement can the project secure enough business-side feedback and buy-in. Depending on the project, this might mean engaging the CEO and twelve VPs—or it may only take the right individual line-of-business manager who oversees pertinent operations. Either way, the core team must break barriers and forge a prosperous, two-way biz/tech exchange.

Learning the Hard Way

> We believe the two major reasons [for non-deployment] are the absence of strong leadership and a lack of buy-in . . . you conduct it without involving the stakeholders and then upon completion tell them, "Here's what the data shows. Now make use of it."
>
> —Jeff Deal and Gerhard Pilcher, *Mining Your Own Business*

The gay.com project was doomed to fail. I was a one-man show: the project champion, vendor, data wrangler, and data scientist all in one. I lacked the proper leadership practice, so, despite all my ambition, enthusiasm, capabilities, and best efforts, I couldn't convince my client to embrace my model and launch a new operative process to leverage it.

Like so many modeling projects, it came down to this dialogue:

Data scientist: I developed the predictive model and the performance is great!

Client: That sounds interesting.

Data scientist: Look at the projected ROI, if only you'd act on it.

Client: What do you mean?

Data scientist: Just integrate the model into this new, large-scale operational process.

Client: *(Balking at the audacity of the data nerd)* You want me to do *what?*

This conversation frustrates the data scientist. She throws up her arms and storms away to spend some alone time crunching data. It's bewildering, this seeming lack of vision and unwillingness to seize opportunity.

But, if she sold the project and got it greenlit in the first place, the data scientist holds much of the responsibility. She's plowed forward, executing the technical part of a project that cannot succeed because it's incomplete from a business standpoint.

Many senior data scientists have already come to hold this perspective. They transcend their technical role and provide invaluable leadership. Take Dean Abbott, an industry-leading consultant and the author of *Applied Predictive Analytics*, who holds us quants accountable: "We think, 'Ugh, they just don't get it, they don't understand statistics, they don't understand machine learning.' It's true, they don't. But you probably don't understand business and all the inputs they're getting, the stressors they have, either. So, we have to lower the wall, be humble, ask a lot of questions, let them explain things in their language, and then try to translate the language so you can be on board with the same ideas."

When I pitched gay.com, I hadn't yet joined the enlightened minority. To help the company hold on to more customers, I should have sold them on just that—not ML (or *predictive analytics*, as was the more common term back then). Rather than churn modeling itself, I should have reframed my pitch to focus on a new operational effort, with a parenthetical mention of the technology that will drive it:

> **We will run a new marketing process to retain customers (using ML to target those most at risk of canceling).**

Gay.com may not have bought in, but at least I wouldn't have sold them on a technical project fated to require as-yet unrecognized operational efforts before it ever realized value.

Learning the Easy Way

Even after my experience with gay.com, I still hadn't learned my lesson. A single experience was not enough for me to see a pattern. Besides, I'd been paid well for my time, so I wasn't tremendously motivated to evolve.

But there are two ways for your second ML project to succeed. One is by experiencing a reckoning that leads you to intentionally follow a well-honed enterprise practice. The other is by dumb luck.

When I landed my second gig as an ML consultant, I got lucky. Sometimes, you learn your lesson the easy way rather than the hard. The next chapter kicks off with that story.

1 Value

Establish the Deployment Goal

In the first step of a machine learning project, you establish the value proposition: what the model will predict and how those predictions will improve operations. This is the deployment goal, the launch intended to take place as the final project step. Your backward planning has begun.

As with any technology, an ML project must begin with a clear business objective, such as "decrease miles driven" or "retain more customers." But that's not where participation from business stakeholders ends—it's only where it begins. Business considerations also inform several semi-technical aspects of precisely how that objective will be pursued by model deployment.

Beyond helping with the project's execution, bringing business stakeholders into the details also helps greenlight the project in the first place. Since the model is meant to seize some degree of control, those in charge must be willing to accept some loss of control. To achieve their buy-in, you must offer full transparency by ramping them up on the concrete way predictions will drive operational decisions. This way, they will come to embrace not only probabilistic thinking, but probabilistic *doing*.

The BizML Practice:

1. *Value*: Establish the deployment goal.

2. *Target*: Establish the prediction goal.

3. *Performance*: Establish the evaluation metrics.

4. *Fuel*: Prepare the data.

5. *Algorithm*: Train the model.

6. *Launch*: Deploy the model.

As the editor in chief of *ITworld*, Jodie Naze pronounced advertising "the lifeblood of the Internet." Without it, she wrote, "the Internet would still be an isolated plaything for the sole use of the academic elite."

When I took a call from the company that was to be my second client, they asked me to use machine learning to better target ads—the very ads that provided their revenue. This company was in the sure-fire business of giving away money: It was the leading search engine for student grants and scholarships. One in three college-bound high school seniors signed up to find out what kind of financial aid they were eligible for. In this book, I'll call the company *EduPay*.

Users "paid" to use EduPay in the same way you pay for most online content: by seeing ads. But these ads were different. First, most were relevant to a student's interests. Some presented ways to cover tuition, such as student loans and military recruitment. Others pitched universities. Moreover, EduPay visually integrated the ads into the overall user experience. Some ads would even ask for a bit of info such as a phone number or intended study year in exchange for connecting the user to the sponsor—rather than just begging to be clicked on. As a result, many users would not even notice that they were interacting with a paid commercial.

But there was room for improvement. The current EduPay system didn't personalize ads. It selected them based on overall popularity rather than relevancy, prioritizing those with the highest response rates across users. This is a solid and common tactic, but it doesn't select ads based on the user's individual preferences.

Personalizing ads with ML nudges up the number of ads that users experience as informative. At EduPay, a unique wellspring of data amplified ML's potential to personalize. Their site collected rich profiles when users registered, including their backgrounds, interests, and educational plans. This presented just about as good an opportunity to target ads as you could hope for.

But Melissa, the director of advertising products at EduPay, didn't call me only to improve the user experience. She wanted to make the company more money. The advertisers paid bounties of up to $25 for each lead. If ad response rates increased, so would the earnings—even without a single new user or advertiser.

The Value Proposition: Defining an ML Application

Let's start at the beginning, by which I mean the end. In planning for a successful deployment, the first thing Melissa at EduPay did right— even before she called me—was to establish the deployment plan, the exact way in which ML would launch. This means specifying what would be predicted and the way in which the predictions would take effect and improve operations.

EduPay's desired business outcome was simple:

> **Desired business outcome**: Increase ad response rates (and therefore revenue).

But for step 1 of an ML project, you must specify how you will get to that outcome, the means to that end. You must declare the way you plan to deploy—first, what will be predicted and, second, what will be done about it:

> **Application**: Ad targeting
>
> 1. *What's predicted*: Will the user respond to this ad?
> 2. *What's done about it*: Display the ad to which the user is most likely to respond.

Since these two ingredients describe the way in which you're applying ML, they specify the *ML application*, a.k.a. the *ML use case*. They establish the *value proposition* for a given ML project. The two determine, respectively, what we will do during the two culminating steps: step 5, train the model to predict something, and step 6, deploy the model into the field to drive operations.

Backward Planning: Forging a Path to ML Deployment

AI strategies fail because AI is a means, not an end. "Do you have an AI strategy?" makes as much sense as asking, "Do we have an Excel strategy?"
—Mihnea Moldoveanu, management professor, University of Toronto

Being AI first means using it last.
—Will Grannis, founder and leader, Google Cloud's CTO office

Focus on decisions and work backward.
—Jack Levis

The EduPay project stood a chance of successfully launching—because we had a precise definition, from the get go, of how ML models would actively improve operations. Every ML initiative must establish this up front, at its inception. Many fail to do this and therefore fail, full stop.

By establishing the end goal, you have a destination that keeps your project moving in the right direction. Rather than using advanced tech for its own sake, you're pursuing an operational purpose.

All planning is backward planning. You start with a goal and work out how you're going to get there. Say you're writing a movie script. According to Hollywood screenwriter Steven Pressfield, "Start at the end. Begin with the climax, then work backward to the beginning. *Carrie. The Great Gatsby. Thelma and Louise.* The ending dictates the beginning. I'm a huge fan of this back-to-front method. It works for anything—novels, plays, new business pitches, music albums, choreography. First, figure out where you want to finish. Then work backward to set up everything you need to get you there."

There's a growing consensus that we've been getting ML backward by not planning it backward. The problem should come first, not the technology. Folks are pushing back on the inverted notion of *The AI-First Company*—a term coined by Alphabet CEO Sundar Pichai and consecrated as a book title by famed venture capitalist Ash Fontana.

While prioritizing ML certainly has merit—as proven by Alphabet itself, given that Google powers Internet search with ML—"AI-first" as a movement tends to suggest first adopting the technology and only later determining its specific use. In this way, it treats AI as a silver bullet sure to bolster enterprise functions.

This is a type of *technology-first* or *solution-first thinking* (here, "solution" refers to the technology itself, not the effective use of a technology). Those who oppose that thinking suggest that you "love the problem, not the solution," as Ash Maurya put it in his book *Scaling Lean*.

"AI can't come 'first,'" wrote ML thought leader and executive Richard Heimann in a review of *The AI-First Company* for the publication that I edit, the *Machine Learning Times*. "If being AI-first literally means solution-first, we lack problem-specific information required to know anything about the right solution and value propositions. We will also lack customer- or market-relevant direction and fail to align strategy with the business."

MIT Technology Review asked ML industry leader Andrew Ng how he responds when people ask him, "How do I build an AI-first business?" Ng didn't take the bait. "I usually say, 'Don't do that.'" He continued, "If I go to a team and say, 'Hey, everyone, please be AI-first,' that tends to focus the team on technology, which might be great for a research lab. But in terms of how I execute the business, I tend to be customer-led or mission-led, almost never technology-led."

It's not only thought leaders. Data scientists in the trenches also warn others about this hard-learned lesson. When a senior ML engineer at Bolt named Francesco Pochetti weighed in online, his usual two or three likes per post jumped to over 1,700, plus scores of reposts. Here's what he tweeted:

Francesco Pochetti
@Fra_Pochetti

All ML projects which turned into a disaster in my career have a single common point:

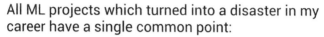 I didn't understand the business context first, got over-excited about the tech, and jumped into coding too early.

10:08 AM - Mar 12, 2022

 Read the full conversation
♡ 1.7k   287 Retweets

And yet, if you prudently eschew "AI-first" and start by establishing the business use case, you've only just begun. You're reading an entire book rather than a bumper sticker for good reason: In and of itself, the industry's helpful change in tone isn't enough. This chapter's step of establishing the deployment goal is necessary but not sufficient. It's only the first of bizML's six steps. The remaining steps are also needed to achieve real operational change.

In fact, this chapter alone won't even fully prepare you for step 1—only the entire book will. Since this first step of the project is to rigorously plan for the final step, deployment, you can't complete the first step until you're familiar with the deployment concepts covered in that culminating chapter. The chapter you're now reading introduces the need for backward planning, outlines step 1, and surveys your business-application options—but chapter 6 is where you'll find the logical and mechanical particulars of model deployment that must be established in advance. As you'll see, the salient particulars aren't arcane or "in the weeds" technically, but that chapter will inform just how detailed this first step's deployment plan must become—more detailed than, "Display the ad to which the user is most likely to respond." I wrote this book to be read front-to-back, but then quite possibly referenced in reverse.

Why Deployment Requires a Mental Leap

Paradigm shifts don't come easy. To deploy ML is to run mammoth numbers of decisions probabilistically, to systematically apply probabilities on a much lower level of granularity than traditionally conceived of. If you're a data scientist, allow me to shatter your world: Probabilistic thinking is a trendy notion, yes, but probabilistic *doing*—methodically acting on probabilities—is barely on the radar for most people.

In most popular media and books, probability serves only to support singular decisions, rather than millions of everyday operational ones. Nate Silver's popular book, *The Signal and the Noise*, tells us to "think probabilistically." It focuses on singular events such as terrorist attacks, economic recessions, pandemics, and political elections. And Peter Bernstein's bestseller *Against the Gods* chronicles the historical development of probabilistic methods. Neither book even hints at applying probability to change the very way we drive our *large-scale* operations. It's one thing to think probabilistically—it's another to *act* probabilistically hundreds of thousands of times a day.

The first deployment takes a mental leap. Even rolling out the most well-trodden application requires a revolutionary mindset, if it's your organization's first time—which is today still the case for most applications at most organizations, as ML's commercial deployment is still relatively young. For example, imagine a company that's considering deploying ML for fraud detection for the first time:

Application: Fraud detection

Desired business outcome: Catch more fraudulent transactions.

1. *What's predicted*: Will the transaction turn out to be unauthorized?

2. *What's done about it*: Manually audit the transaction.

Before deploying ML, a team of human auditors is doing their best to comb through transactions to find those enacted by criminals.

These mere mortals can only examine a small fraction, so inevitably they employ ad hoc heuristics to focus their endless search. Targeting auditor activities with a model would add some science to this ad hoc method. It would formalize and improve the choice of which transactions auditors spend time on.

To deploy the model, management must accept a loss of control. The model takes the reins, determining a slew of transactions that will never be audited: those scored as improbable by the model. Any fraud therein will go undetected. And yet, if management forgoes the model, succumbing to the temptation to check out those unflagged cases, they're only compromising the benefit of model deployment. The whole point of the model is to flag cases worth auditing.

The key to getting decision-makers to not only authorize this leap but collaborate on it is bringing them up to speed on the complete project plan. Trust and comfort will grow as they become familiar with the deployment goal and the steps to get there. This means ramping them up in detail, including making clear what outcome the model will predict, the performance metric for evaluating that model, the source of data, the data science resources that will be needed for producing a model, and the engineering plan for its deployment.

The means to accomplish this ramp-up and full project specification? *Conduct the remaining five bizML steps that come after this one— collaboratively, with business stakeholders.* In so doing, you will enlist them to understand, weigh in on, and approve the project's particulars.

But before we proceed with the rest of the steps, let's finish this one first. It's time to pick your value-driven deployment goal for ML.

Decisions, Decisions: Picking Your First ML Project

The analytics is the easy part. We've nailed that. The first, more challenging part of a machine learning project is to fully define the business problem. If you take the time up front to get that right, then, at the end of the day, you'll be directionally in the right place, making it a lot faster and a lot less costly than having to make course corrections toward the later, more expensive parts of the project.

—Gerhard Pilcher, president and CEO, Elder Research

Home in on the ML application that offers the greatest potential impact. Pick the lowest-hanging fruit: the application that improves the large-scale operational process with the most room for improvement.

And yet, start small. Even if the potential win is big, the project's scope is finite in one key way: You only solve one problem at a time. Each project must focus on a single, modestly scoped opportunity with laser precision. This means settling on one application of ML, a single value proposition defined by what's predicted and what's done about it. This increases the likelihood that your organization's first ML initiative will succeed. That early win will set a valuable precedent for future ML projects.

Choosing which process to improve with ML is a business decision very much specific to each company. There's no one-size-fits-all answer. It depends on your sector, your business model, the nature of your operations, and how much they could stand to improve. Your current operational model, method, and culture must guide the choice.

First and foremost, spend time with operational stakeholders to understand their pain points and constraints. As a group, the stakeholders constitute your client for the project. That's the customer you are here to serve.

Explore. What's the ripest high-volume operation, the process performed repeatedly many times at your organization that could most stand to benefit first? Is there a frequent operational decision that's already entirely or partially automated but not yet optimized?

The options are seemingly endless. Predictive models determine which tax returns to audit, which customers to contact for marketing, which debtors to approve for increased credit limits, which patients to clinically screen, which employees to woo away from quitting, which persons of interest to investigate, and which equipment to inspect for impending failure.

As you investigate opportunities and weigh your options, let's get a lay of the land by surveying some of the more common ML applications.

The Choice Depends Partly on Sector

The types of use cases with the greatest value potential vary by sector. . . . In consumer-facing industries such as retail, for example, marketing and sales is the area with the most value. In industries such as advanced manufacturing, in which operational performance drives corporate performance, the greatest potential is in supply chain, logistics, and manufacturing.

—McKinsey's "Notes from the AI Frontier"

Some of ML's more typical use cases apply across industry verticals, such as those in marketing and advertising. Others are industry-specific. Here are some examples:

Application and desired business outcome	What's predicted (model output)	What's done about it (deployment)
Response modeling to increase the marketing response rate	Will the customer buy if contacted?	Mail a brochure to those likely to buy.
Targeting ads to increase clicks	Will the user respond to this ad?	Display the ad to which the user is most likely to respond.
Churn modeling to decrease customer attrition	Will the customer defect if not contacted?	Reach out with a retention offer to those most likely to defect.
Credit scoring to decrease defaults	Will the debtor default on their loan?	Deny risky applications for credit.
Supply chain management to optimize inventory	How much demand will there be for each item?	Maintain stock levels accordingly.
Delivery prediction to plan for more efficient delivery	Will the address receive a package delivery?	Plan the delivery truck assignments of predicted packages alongside known ones.

In marketing, response modeling is the most established application, with several decades of proven results. But, as a rule of thumb, churn modeling is arguably a hotter marketing application since it's often a more cost-effective way to grow business. Despite that, your circumstances may veer away from this trend. For example, if your customer attrition rate is very low, that limits the potential gain with churn modeling, so response modeling may be a better fit.

In financial services, credit scoring is common practice. If your company isn't already applying ML to assess the risk of each debtor, there could be a tremendous opportunity to do so. Insurance companies commonly apply ML to set prices and fast-track claims. On the other hand, most mid-sized and large companies could benefit from targeting their marketing, including financial services firms. The greatest opportunity with ML is not dictated by the sector alone.

In the public sector, ML is often applied to manage risk, targeting all kinds of safety measures and proactive investigations for risky situations such as bridges that risk collapse, buildings more likely to catch fire, manholes more likely to explode, homes more likely to expose kids to lead poisoning, restaurants more likely to violate health codes, drivers to flag as less vigilant, and workplace practices more likely to lead to injury.

Detecting a Situation Instead of Predicting an Outcome

For some ML applications, the model detects or diagnoses a problem rather than predicting the future—although by convention we use the word "predict" anyway, such as when "predicting which transaction is fraudulent," and we still call it a *predictive model*. In practice, it's largely the same kind of endeavor. As with models that predict a future outcome, a "detection" model is attempting to ascertain an unknown and the same technical modeling methods apply. Here are some examples:

Application and desired business outcome	What's predicted, i.e., detected (model output)	What's done about it (deployment)
Fraud detection to prevent more fraud	Is the transaction fraudulent?	Place a hold on high-risk transactions and/or send them to human auditors.
Healthcare diagnosis to improve treatment	Does the patient have the condition?	Flag the patient for additional tests to potentially confirm the diagnosis.
Spam filtering so you see less spam	Is the email message spam?	Relegate spam to a separate email folder.
Speech recognition to transcribe the spoken word	Is X the word that corresponds with the audio segment?	Label the segment with the word predicted as most likely.
Fault detection to decrease the number of broken items	Is the item faulty (e.g., as it rolls off a factory assembly line)?	Inspect items predicted as likely to be faulty.
Autonomous driving to lessen human workloads and improve safety	Is there a stop sign in the image?	Bring the vehicle to a stop when a stop sign is detected.

There are many, many more ways to apply ML. Don't be afraid to invent a new one, according to the exact nature of your business and its main operations. Allow the wide range of established use cases to potentially inspire an original one. New ones crop up all the time. ML has been used to sort cucumbers, predict the outcome of construction projects, detect online trolls and sewage leaks, and catch scooters being illegally ridden on the sidewalk.

I would like to encourage you to also consider social-good applications that benefit society. For example, models can identify those most vulnerable or at risk in order to intervene. The nonprofit Predict Align Prevent applies ML to identify children at risk for maltreatment. The organization's models improve early detection and intervention. As another example, the US Equal Employment Opportunity Commission predicts discrimination, flagging which groups of people in a specific industry are more susceptible. ML also serves as an important tool for climate technology by, for example, predicting the amount of carbon captured by reforestation and regenerative agriculture projects, predicting supply and demand for renewable energy and storage, and predicting extreme weather risks to real estate and crops (the conference series I founded includes an offshoot focused on this, Predictive Analytics World for Climate Technology).

By deciding on an ML application, you've established your deployment goal. Now all you've got to do is sell it.

Getting the Green Light

This book ends with the project's beginning—the conclusion guides you on how to pitch ML projects. But for now, let's wrap up step 1 with how UPS and EduPay gained traction for their respective use cases.

When Jack Levis wanted to optimize UPS delivery routes, he had a lot of convincing to do. "The calls I got were incredible. 'It's time to stop this. No computer can tell a professional a better way to deliver. You've gotta get your head out of the clouds.'"

But Jack had an ace up his sleeve: He was selling potential gains, not just cool technology. And he knew that the key to convincing was disclosing the details of the intended deployment and even demonstrating it in action—literally, as when he took the executive Chuck on that fateful ride.

Jack also knew to enlist while he convinced. As an ML project progresses, selling and educating evolve into collaborating. The stakeholders and line-of-business managers should transform from skeptics to key team members. They run the operations, so they should help inform the changes to those operations. You can't hash out a fully detailed deployment plan without them. As you indoctrinate, also recruit.

In contrast, when it came to selling operational change, EduPay was at a disadvantage: They had me rather than Jack Levis. I had emerged from academia in love with the tech. And even after the disappointment of the gay.com project, I hadn't yet learned the lesson: The deployment plan must be sold first.

Fortunately, I wasn't in charge. I was serving as the data scientist, not the project leader. EduPay had come to me with a specific deployment plan already in mind. My job was to devise a technical approach to modeling. I didn't have to sell EduPay on operational change. The company was already set to do so.

The project leader, Melissa, had the wind at her back. After conceiving this project, her brainchild, she'd already secured the approval of her superiors to implement her vision. This has gone smoothly. EduPay was a small company and Melissa had a great deal of autonomy in her ownership of the ad system. Moreover, the change she proposed was relatively incremental: Instead of automating ad-selection decisions for the first time, her goal was merely to change how those automatic decisions were being made. EduPay's system was already selecting which ad to display, based mostly on each ad's overall popularity—now we were going to change it to select which ad to display for each individual user.

Next Steps

But even with a path to deployment relatively unobstructed, I wasn't ready to train any models for EduPay just yet. Deciding on the value proposition—what's predicted and what's done about it—is only the first pre-production planning step. The next step is to get more specific about what the model will predict—a lot more specific.

As we'll see over the course of the remaining chapters, there was still much to learn from the EduPay project. We may not all be natural-born ML leaders like Jack Levis, but we can learn to do it right.

2 Target

Establish the Prediction Goal

As we discussed in the previous chapter, to plan for machine learning deployment, you must begin by establishing what will be predicted by the model and what will be done about each prediction. In this chapter's step, you more fully define the first of these two, the prediction goal. It must be specified in great detail.

Welcome to a key intersection between tech and biz, where business pragmatics inform the semi-technical details and where business stakeholders must delve into these details. Your mission, should you choose to accept it, is to forge a rare collaboration, enlisting business leaders to weigh in on the caveats and qualifications that determine the prediction goal in all its detailed glory. If you succeed, you will have translated a broadly defined business intention into a well-defined requirement for technical execution.

The BizML Practice:

1. *Value*: Establish the deployment goal.
2. *Target*: Establish the prediction goal.
3. *Performance*: Establish the evaluation metrics.
4. *Fuel*: Prepare the data.
5. *Algorithm*: Train the model.
6. *Launch*: Deploy the model.

Jack Levis doesn't remember the exact moment of epiphany when delivery prediction came to him. It's the kind of great idea that just seems like a given once it's on the table. But before anyone had conceived of it, UPS couldn't fully optimize. Recall how Jack formulated the paradox that had been holding the company back:

> **The Delivery Paradox:** You can't optimally plan the truck-loading until you know all the deliveries that will need to be made. But by the time you know all the deliveries, you've run out of time to load the trucks.

You can only predict your way out of a paradox like this. The same kind of conundrum arises for optimizing all kinds of logistics, whether you're carpooling a bunch of kids, any one of whom may call in sick at the last minute; you're casting big stars in a movie, any one of whom could surprisingly say yes just after you've offered the role to someone less famous; or you're accepting a new consulting client, even though you know that a bigger client opportunity could come tomorrow. But these operations are too small for machine learning. Let's be real: Your kid's carpool probably wouldn't benefit.

When applied to large operations, predictions help, even without unrealistic expectations as to their precision. When UPS predicts a delivery, it's presumed without being presumptuous. It's only tentatively treated as a given—just for the purposes of loading the trucks. By the time the trucks roll out, all the packages have become real—they've arrived at the shipping center and wound up on a truck. Any predictions that didn't materialize are discarded.

With such a well-conceived scheme—powerful yet prudent—you might expect Jack to have triggered the ML project ASAP. After all, the prediction goal appears clear and simple: *Predict which addresses will receive a UPS delivery tomorrow.* So crank up the predictive engine and pour in the data.

Not so fast. Jack knew that the devil lurked in the details. Should the model predict which building would receive a package or should it be more granular, predicting for individual apartments or business suites within a building? Should it predict the number of packages or just whether there'd be at least one stop needed at the building? And how about breaking down predictions by the time of day? Some packages must be delivered first thing in the morning, while others are promised delivery only by the end of the day. At the time, UPS offered thirteen levels of service that guaranteed delivery by different times of day.

The Deployment Plan Informs the Prediction Goal

If ML seems cool but these kinds of details seem boring, then you're suffering from a common but treatable condition. Here's the antidote, a law that keeps your focus on value:

> **The Law of ML Planning**: Keep deployment at the top of your mind. The precise way in which predictions will affect and improve operations informs every step of the ML project.

With faithfulness to that law, Jack looked hard at how the predictions would be used. Every day, UPS's existing optimization process assigned each truck to a region that could be handled by the driver within one workday. Several factors determined whether an assigned workday would be manageable, including how large the region was, the number of stops needed, and the number of packages at each stop.

This procedure functioned everywhere, from dense urban centers to rural areas where stops were separated by miles. It handled this diversity with dynamic granularities. Out in the country, a truck could be assigned a few zip codes for tomorrow's outing. But in the city, a truck might cover an area that included only a couple dozen delivery addresses.

For package prediction to be compatible with the current procedures, it had to speak the same language. It had to make predictions at the same varying levels of granularity, depending on the region. At UPS, this kind of unit that varies in size is known as a *sequence*. It's the smallest unit of geography that is taken into consideration when planning a truck's packing and deliveries. In some areas, a sequence corresponds to a single delivery address. In less populous areas, it may correspond to a group of addresses. And in the most rural regions, a sequence corresponds with an entire zip code.

Since the existing planning process operated in terms of each package's sequence, *predicted* packages would also need to be determined on the sequence level. But it didn't quite stop there. Within a sequence—whether large or small—the number of packages and the number of times the truck would need to make a stop were critical to determine whether a driver's assignments were doable within one workday. After working through the details, here's how Jack formulated his prediction goal.

> **UPS Prediction Goal:** For each sequence (a kind of dynamic geographical region), how many packages across how many stops will be required tomorrow—for each delivery time commitment? For example, sequence 457, a group of three office buildings with twenty-four business suites, will require two stops with three packages each by 8:30 a.m. and five stops with eight packages each by 5:00 p.m.

Before UPS introduced package prediction, the planning process already worked with this level of detail for known deliveries. Once this process could also work with the same level of detail for *predicted* deliveries, it could plan even more effectively. With the predictions designed to be compatible with the existing process, they fit right in.

Pursuing this prediction goal took some doing. The core model predicted delivery probabilities on the individual delivery address level. But the prediction goal generally demanded predictions for a broader geographical area—plus, each prediction was not to be a yes/no "will there be a delivery?" prediction, but instead a prediction of *how many*. To calculate this quantitative prediction for each sequence, some hand-designed code "rolled up" individual per-address predictions, aggregating them into the required per-sequence prediction.

The Difference between ML Failure and Success

It turns out that it's easy to mess this up, failing to align the prediction goal with how predictions will be operationalized. Since this is only step 2, such misalignments throw off the project early on. Defining the prediction goal marks a crucial turning point that all ML projects hit soon after their conception. Many projects get it wrong—they neglect to scrupulously flesh out the prediction goal before jumping into the analysis. This is a deadly error, usually resulting in a model that's embalmed rather than embraced. So properly defining the prediction goal is your chance to shine—not by flaunting advanced methods but by exercising restraint and meticulousness.

Let's revisit the classic marketing application, response modeling. It aims to increase profit by targeting marketing. Here's how we described it in the previous chapter:

Application and desired business outcome	**What's predicted** (model output)	**What's done about it** (deployment)
Response modeling to increase the marketing response rate	Will the customer buy if contacted?	Mail a brochure to those likely to buy.

As we see here once again, each ML application is defined by *what's predicted* and *what's done about it*. In this bizML step, we're defining the prediction goal—the "what's predicted" part. In a later step, data preparation will use that prediction goal to populate a column of data, which is known to data scientists as the *dependent variable* and in this book is called the *output variable*. With this column in place, the overall dataset is referred to as *labeled* or *supervised* data. After preparing it, an ML algorithm will use the data to train a model that predicts as well as possible. With the current step, we are defining the goal for that predictive model.

The prediction goal has got to be more specific. We must fully define it, in all its detailed glory. For most any response modeling project,

"Will the customer buy if contacted?" leaves too much unspecified. While we iron out the details, the Law of ML Planning tells us to keep deployment at the top of mind. For this application, the model's predictions will target marketing. They will determine who is and who is not included in a campaign's contact list.

The question then is, Who's worth marketing to? Well, a customer must buy enough to be profitable. And we need a realistic time frame—a customer who buys a year later doesn't normally count for campaign ROI, and that kind of lag makes tracking difficult. With that in mind, here's a step in the right direction:

> **Response Modeling Prediction Goal (hypothetical example):**
> If sent a brochure, will the customer buy within thirteen business days with a purchase value of at least $125 after shipping and not return the product for a refund within forty-five days?

Now we're in business. We've fully defined the prediction task that ML will pursue, the requirements specifications for the predictive model. And it's based on the particular business context: the existing practices into which the model's predictions will integrate and the way they'll take effect. This is the level of detail you need to establish before preparing the data and training a model with it.

Prediction goals like this determine what a model will do for each individual. The goal poses a question about a single individual—a yes/no question for *binary models* (see the following sidebar). That's the question the model will attempt to answer each time it scores an individual. "Will *the* customer buy?" refers to the one individual that a model is scoring, since it scores only one at a time.

Before you dive too deeply into defining a prediction goal, there's a fundamental pitfall to consider: *Settling on a poor approximation of the true prediction goal.* Sometimes, the available data is only a proxy for the pertinent target of prediction. For example, law enforcement aims to predict future crime to drive sentencing and parole decisions, but only rearrest data is available. The *ground truth* of whether a released defendant went on to commit a crime is not directly known—it's only approximated, imprecisely, with data tracking whether they were arrested again. But using "Will the convict be rearrested?" as the prediction goal brings up a central ethical issue with *predictive policing*. Because historically disadvantaged groups such as Black Americans are more heavily policed, this unfairly inflates the relative frequency of arrests and, as a result, of predicted arrests by a model trained for that prediction goal. Similarly, as another example, a widely used model for guiding healthcare treatment was trained to predict healthcare *cost* rather than *need*. Since the data reflected that less had been spent on Black patients on average, the model was shown to disadvantage those patients. This book's conclusion surveys other ethical issues that arise with ML's deployment.

Binary Models

Unlike UPS's goal for package prediction, response modeling pursues a *binary* prediction goal—it predicts the answer to a yes/no question. This holds for both for the simplified goal, "Will the customer buy if contacted?" as well as the more complete prediction goal just discussed—and it holds also for the remaining examples in this chapter and most of this book. A model trained for a binary prediction goal is called a *binary model* or a *binary classifier*.

For most new ML initiatives, binary models are usually the best place to start. Most any prediction problem can be framed as binary—to classify which individuals do or do not behave in some defined way, such as whether they spend beyond a certain amount, default on their credit, click, buy, lie, or die. This applies for detection tasks as well: Is this transaction fraudulent? Is this email message spam? or Is this manufactured item faulty?

But models can also predict *how much* or *how many*, rather than only *whether*. Such models predict the number of purchases, amount expended, magnitude of insurance claims, or a customer's *lifetime value*, which is the amount of revenue or profit expected across a relatively long period of time (an ideal yet sometimes overly ambitious prediction goal). These are sometimes called *continuous* or *numerical* models.

Binary models hold a couple of significant advantages. First, even though they pertain to only two possible outcomes, they estimate where that outcome falls on a continuum of likelihoods. That is, they provide a probability for the outcome—*how likely is it the customer will buy?*—rather than only outputting a definitive "yes" or "no." Since a probability relays the degree of uncertainty, it more effectively drives all kinds of decision-making processes, from pricing to ad selection to risk management.

Further, models that predict "how much" tend to be more technically challenging—both to develop and to evaluate. For many projects, binary models offer a more straightforward approach that avoids introducing unnecessary complexity.

Proactively Preventing Bad Outcomes

Some models aim to predict bad outcomes, in order to intervene before they come to fruition. For marketing, this means predicting customer defection so that we can target retention efforts most effectively. Enter *churn modeling*:

Application and desired business outcome	What's predicted (model output)	What's done about it (deployment)
Churn modeling to decrease customer attrition	Will the customer defect if not contacted?	Reach out with a retention offer to those most likely to defect.

But just saying "Let's predict which customer will defect" omits some crucial details. Does the question pertain to all customers, even brand-new ones? Are we only predicting explicit quitters, or do we also want to predict those who significantly decrease their business with us? And how far ahead are we predicting?

Even back at my first consulting gig with gay.com, I got this part right. By naively selling the modeling piece rather than genuine operational change, I'd killed my chances of achieving deployment. But I did recognize the need to carefully define the prediction goal so that the model's outputs could be actionable in a business context:

Churn Modeling Prediction Goal (for gay.com): For customers on monthly and quarterly plans only, will the customer intentionally and explicitly cancel within three months? Passively canceling due to a credit card failure doesn't count.

The details matter. We want to predict would-be cancelers who we might still be able to save. Predicting too far into the future, such as a year from now, often proves too difficult for a model. Besides, we don't want to reach out with an incentive to stay—such as a costly discount offer—until their departure is more imminent. This means setting aside those customers on annual plans, since their next renewal may be up to a year away (they could be handled with a separate churn modeling project). Finally, we can only try to change the mind of those who will cancel on purpose. For those whose credit card charge is declined, which leads to an automatic cancellation, it may suffice for the company to take reactive rather than predictive action, with potentially no need to offer a discount. Those customers aren't canceling intentionally, so their attrition is another matter for another project.

Predicting Too Late

Some churn modeling projects predict well but predict too late. ML consultant Karl Rexer shared with me a perfect example of this that arose when he was helping a bank. The goal was to identify customers likely to close their accounts. Karl discovered that if the customer's deposit and loan balances fell very low—in combination with some other factors about the customer—they had a 44 percent chance of closing within a month. On the scale of things, that's an astounding find. Relatively speaking, 44 percent is very high. Across all customers, the chances were only 2 percent.

For any one of these customers, the bank faced a high risk of losing them. Unfortunately, as Karl realized, there wasn't much they could do about it. These predictions were almost useless since it was already too late to change their mind. That is to say, the insight wasn't *actionable.* By the time the account holder lowered their balances, they already had one foot out the door. They'd already largely emptied their accounts and wound things down, and only had to formally close out their account as an administrative step. By this point, such customers prove almost impossible to save.

For these predictions to be valuable, they'd need to be made further ahead. If a customer is destined to leave, but not for several months, there's still a chance they could be saved. It's harder to predict further ahead, so you may not get as glorious a predictive boost, but, in doing so, the predictions are often much more actionable. After all, there's more time to intervene if the customer's likely cancellation is further out in the future at the time that the customer is flagged as a high risk. And, as we'll see in the next chapter, even when a model produces less confident odds, it can still be highly valuable.

This is what Karl did with his banking client—in combination with another tactic: Rather than predicting whether a customer would fully close their accounts, his model predicted whether the customer would partially disengage by exhibiting a steep drop in transactions. This made the predictions more valuable. By predicting earlier behavior that often precedes absolute defection, the bank could still take action to potentially prevent losing the customer.

To illustrate what this might look like, consider this plausible prediction goal for an online subscription service:

> **Churn Modeling Prediction Goal (hypothetical example):** Among subscribers who've been around for at least four months, will the customer decrease their monthly usage by 80 percent in the next three months and not increase their usage of another in-house product?

If a customer will drastically reduce their use of the product, that alone is cause for concern, even if we haven't predicted they'll fully close out their account. It would be worth intervening beforehand. Of course, a decrease in usage is only meaningful among customers who've been around long enough to establish a baseline of interaction. This is why we aim to predict only for those with a tenure of at least four months. Finally, if a customer disengages but compensates by increasing their use of another one of our products, this doesn't "count" as defection—that kind of change may not be worth any investment to prevent.

Predicting Intermediate Steps

Predicting a partial step also applies to predicting good outcomes rather than bad—to predicting customer engagement rather than disengagement. Consider targeting fundraising activities for a charitable organization, which is very much akin to response modeling for direct marketing:

Application and desired business outcome	What's predicted (model output)	What's done about it (deployment)
Targeting fundraising to increase donations	Will the individual make a donation?	Mail a letter to those likely to donate.

The standard prediction goal is obvious: Will the individual make a donation? In some cases, this is the best tactic, such as for targeting direct mail solicitations.

But in the world of fundraising, direct mail mostly garners only small donations. Soliciting large donations is a different game. The cultivation of major donors can take years of relationship building. These donations are difficult to predict because they're rare and because it's tough to tie them to any particular prior action. Instead, one charity predicted an intermediate step toward securing a large donation:

> **Fundraising Prediction Goal (a charitable organization):** Will the prospective donor agree to an in-person meeting with a gift officer?

If the prospect is willing to meet, this at least signifies an affinity for the charity and marks the beginning of what will hopefully be a long-term relationship.

This approach worked. By driving email, direct mail, and phone outreach with these predictions, the charity identified 40 percent more high-value prospects willing to meet. And in a retrospective evaluation of the model, the model identified 75 percent of major donors who'd already donated.

Deciding Which Instead of Whether

With the marketing applications we've discussed so far, the model drives the choice between an active and passive *treatment* for each individual: To contact or not to contact—that is the question. Response modeling decides whether to reach out with sales material, and churn modeling decides whether to reach out with a retention offer.

But for EduPay, models needed to decide *which* ad to show. There was no question of *whether* to show an ad—one would always be shown. The passive treatment was never an option. In fact, there were a plethora of options: a pool of 291 ads from which to select. These sponsors had lined up, willing to pay each time a user responded to their ad.

Without ML, the ads were already working very well. By displaying the most popular ads first, the system was already generating $1.5 million in monthly revenue. But the choices were one-size-fits all—kind of like advertising Hollywood blockbuster movies on television. They weren't personalized. There was no attempt to target individual users with the ads they'd be most likely to respond to.

We needed a little trick to define the prediction goal for ad selection, as it isn't straightforward. You can't simply predict, "Which ad would the user respond to?" The reason for this is that we don't have the data. We don't have records tracking any user who was shown *all* the ads to see which they responded to. We never conducted that or any experiment—the aim was to use data that the business had already organically accumulated.

The available data told us how each user responded when shown a single ad—across a limited number of ads for each user. This is the data collected in the regular course of business, a.k.a. *found data*. It encodes the experience from which the modeling process can learn.

The solution to this dilemma was to develop a model for each ad— 291 different models, each predicting whether the user will respond to that ad. This means 291 different prediction goals, each in this form:

> **Prediction Goal for Targeting Ads (EduPay):** Will the user respond to this ad if it is displayed?

To choose an ad for a given user, all the models were systematically applied and the choice of which ad to display was based on the predicted probabilities. Some models might say the chances are only a fraction of a percent, and others might go up to a 20 percent probability or even higher—which is a lot, considering that online ads are more often than not simply ignored. In this way, the choice of ad became personalized, the overall odds improved, and the number of responses increased over time.

Ads weren't only selected based on model probabilities. Other factors were also in play, such as the amount the sponsor was willing to pay and eligibility requirements stipulated by the sponsors. We'll visit those and other mechanics of step 6, *model deployment*, when we get to that step's chapter.

Collaborating on the Prediction Goal

> Engage everyone who will touch the analytic model in the development process. . . . The success or failure of analytics and data science initiatives often hinges on whether those on the "front lines" of business actually use and follow them.
>
> —Tom Warden, Chief Data and Analytics Officer, EMPLOYERS

As you navigate these considerations and formulate the prediction goal in complete detail, you need help from across the enterprise. You can't go it alone. Every semi-technical aspect that defines what you're going to predict depends on pragmatic business considerations. For the resulting model to be operationally viable, you've got to pull together a multidisciplinary team.

By deeply collaborating, business experts and data scientists can hone down the prediction goal to one at the intersection of two sets of prediction objectives, those held by the business side and those offered by the tech side:

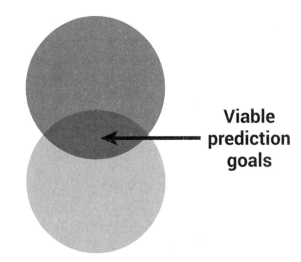

Conceivable prediction goals that would be valuable

Viable prediction goals

Behaviors that can be analytically predicted

On the top, the range of conceivable prediction goals that could be valuable for the business is limited only by the imagination of your operations and marketing staff. They could find value in predicting all kinds of individual outcomes or behaviors, including who will crash their car, fall in love, default on a payment, quit their job, or commit fraud. The problem is, for a given company, only some of these ideas can be feasibly achieved with modeling—depending, for example, on how well it would need to predict and the availability of pertinent data.

The set on the bottom also contains many possibilities. Many things could be analytically predicted, since we have so much data recording so many different kinds of outcomes and behaviors. The problem is, many such ideas that sound appealing in the lab would never actually be used. Of the many outcomes that could be modeled, only a fraction will be business actionable. For example, a data scientist might turn out a model to target a direct marketing campaign, only to find that marketing managers prefer to contact their entire list without targeting any further. Just because the data scientist has created an effective model doesn't mean the business is ready to act on it to drive decisions. All too often, a lack of management buy-in or unforeseen business constraints preclude model deployment. This is the main cause of today's low ML deployment rate.

A shrewd business perspective navigates to a viable prediction goal within the intersection of these two sets—one that's both achievable and usable. You must enlist the wisest operational experts to inform which predicted behaviors hold potential business value. Ultimately, the prediction goal, in full detail, requires insight and buy-in from your collaborators, including those who run the operations that will be affected and driven by model deployment, such as marketing staff, who must be willing to change their targeting accordingly.

By enlisting this hard-core, multidisciplinary collaboration, you're breaking boundaries and engaging in a very rare sort of teamwork. You're involving a larger group in the nitty-gritty, quasi-technical aspects of a prediction goal that has traditionally been below the radar for business stakeholders, such as:

- Whether to predict a final or intermediate outcome
- How far ahead to predict
- Who or what is being predicted—for which kind of customer or other organizational element each prediction will apply

With this sustained business-side involvement, the predictive scores output by your model will deliver the greatest impact. They will be aligned with business strategy and actionable within your company's operational framework. And they will be approved for integration by those managing operations.

Detection versus Prediction: Sometimes Easier, Sometimes Harder

As you define the prediction goal, beware the differences between a goal to predict the future in the literal sense of *prediction* and a goal that predicts a situation or diagnosis—that is, a detection goal. As I pointed out in the last chapter, we do generally use the word "predict" either way, such as when "predicting which transaction is fraudulent," and we still call it a *predictive model*. But there are some significant differences between the two that could wind up making the project alternatively easier or harder.

At first glance, detection goals look easier to define. When it comes to predicting (detecting) which transaction is fraudulent, which email is spam, and which medical image signifies a certain diagnosis, how many caveats and qualifications could there be?

The answer is, not many. Since the model doesn't predict a future event, we need not grapple with the concept of time in defining exactly what's to be predicted. We don't need to decide how far ahead to predict, as in, "Will the customer cancel within three months?" We don't need to define the extent to which the event *happens*, as in, "Will the customer decrease their usage by 80 percent?" And we don't need to consider predicting intermediate actions that occur before the most important behavior, as in, "Will the prospective donor agree to a meeting?"

When defining a detection task, there's not much room for variation. For example, a transaction is either authorized or fraudulent. Specifying the prediction goal tends to be simpler.

On the other hand, projects to predict the future reap the benefits of *big data* in a way detection tasks cannot: Model training can *learn from history*. We know which customers did or did not buy in the past, which canceled their subscription, and which clicked on an ad. This information is already encoded in data that has been collected in the normal course of business. This *found data* serves as experience from which to learn. In step 4, it will form the basis for preparing the *training data* that fuels the pursuit of the prediction goal we're establishing in this step.

In contrast, for detection, the training data must be labeled manually. We don't get to benefit from the "time will tell" nature of found data. Time has told where packages tend to be delivered—and that provides learning examples—but we need humans to label whether a transaction is fraudulent or whether a medical image signifies a positive diagnosis. Because of this, the data needed for training detection models is generally more expensive than the data needed for predicting future outcomes.

And yet the human bottleneck for labeled data isn't always that bad. When we get into step 4—prepare the data—we'll see that, for some detection tasks, although human input is required for each label, we don't need much *additional* human effort beyond what people are already doing anyway in the normal course of commerce.

Model Performance: How Well Can We Achieve the Prediction Goal?

Now that we've defined the functional purpose of a model—what it should do—the next logical question is, How well does it do it? We've established *what* it should predict; now we want to know how *well* it predicts—and which performance metric is the right one to report on. That's the next chapter's step. Warning: If you were thinking "model accuracy" is the way to go, you've got another thing coming.

3   Performance

Establish the Evaluation Metrics

Once you've established *what* machine learning will predict, the next question is *how well* it predicts. Fortunately, evaluating its performance doesn't require becoming a technical expert, since you can benchmark a model without regard to its inner workings. Here, we only judge *how well* it predicts, not *how* it predicts. It's only a matter of arithmetic, not "rocket science."

Often, you will hear of *accuracy*, a simple tally of how often a model predicts correctly. But accuracy is not only the wrong measure for most ML projects; it also feeds a common fallacy that tremendously mismanages expectations.

If not accuracy, then what metric? One is *lift*, a simple multiplier that tells you how many times better than guessing your model predicts. Another is *cost*—the price of each false positive and the (usually very different) price of each false negative.

Once established, the metrics serve to evaluate both model training (step 5) and deployment (step 6). This chapter gets to the heart of the matter. Exactly how valuable is imperfect prediction? In what way do all ML deployments serve to triage and prioritize? And how do you translate from raw predictive performance to true business metrics like profit?

The BizML Practice:

1. *Value*: Establish the deployment goal.
2. *Target*: Establish the prediction goal.
3. *Performance*: Establish the evaluation metrics.
4. *Fuel*: Prepare the data.
5. *Algorithm*: Train the model.
6. *Launch*: Deploy the model.

Headlines about machine learning promise godlike predictive power. Here are four examples:

- *Newsweek*: "AI Can Tell If You're Gay: Artificial Intelligence Predicts Sexuality from One Photo with Startling Accuracy"
- *The Spectator*: "Linguistic Analysis Can Accurately Predict Psychosis"
- *The Daily Mail*: "AI-Powered Scans Can Identify People at Risk of a Fatal Heart Attack almost a Decade in Advance . . . with 90% Accuracy"
- *The Next Web*: "This Scary AI Has Learned How to Pick Out Criminals by Their Faces"

It's all a lie. ML can't confidently tell such things about each individual. In most cases, these things are too difficult to predict with certainty.

Here's how the lie works. First, researchers report high "accuracy," thereby suggesting—and reliably misleading the reader to believe—that *their model can identify both positive and negative cases and generally be right about it either way*. For many prediction problems, that level of performance is achievable only in science fiction.

These reports later reveal—buried within the details of a technical paper—that they were misusing the word *accuracy* to mean another measure of performance related to accuracy but in actuality not nearly as impressive as "high accuracy" implies.

But the press runs with it. Time and again, this scheme succeeds in hoodwinking the media, a beast that all too often thrives on hyperbole. This time-honored tactic repeatedly generates flagrant publicity stunts that mislead.

Now, the predictive models they're reporting on often do deserve high praise. The ability to predict better than random guessing, even if not with high confidence for many cases, improves all kinds of business processes. That's paydirt. And, in some limited arenas, ML *can* predict extremely well, such as for recognizing objects like traffic lights within photographs or recognizing the presence of certain diseases from medical images.

But many human behaviors defy reliable prediction. Predicting them is like trying to predict the weather many weeks in advance. There's no achieving consistently high certainty. There's no magic crystal ball.

Stanford's "Gaydar" Doesn't Perform at Face Value

Take the hype surrounding Stanford University's infamous "gaydar" study. In its opening summary (the abstract), a paper published in 2018 by researchers Michal Kosinski and Yilun Wang claims their predictive model achieves 91 percent accuracy in distinguishing between gay and straight males from facial images. This inspired journalists to broadcast grossly exaggerated claims of predictive performance. One *Newsweek* article kicked off with, "AI can now tell whether you are gay or straight simply by analyzing a picture of your face." The front cover of the *Economist* depicted a face looking like a fingerprint along with, "What machines can tell from your face."

This resulting deceptive media coverage is to be expected. The researchers' opening claim of 91 percent accuracy tacitly and inevitably conveys—to lay readers, non-technical journalists, and even casual technical readers—that *the system can tell who's gay and who isn't and generally be correct about it for both categories.*

But that assertion is false. The model can't confidently "tell" for any given individual in general. Instead, what Stanford's model can do 91 percent of the time is much less remarkable: It can identify which of a pair of two males is gay *when it's already been established that one is and one is not.*

This benchmark—which I call the *pairing test*—can sound like a compelling story, but it's a deceptive one. At first, it may look like a reasonable indication of a predictive model's performance, since the test creates a level playing field where each case has 50/50 odds. And, indeed, the result of this test does confirm that the model performs better than random guessing. Most data scientists know the pairing test by a more technical name, *AUC* (Area Under the receiver operating characteristic Curve). And yet most data scientists, in my experience, haven't come to realize that the two metrics are one and the same. AUC is mathematically equal to the performance observed running the pairing test (assuming you run it enough times). They're just two different ways to estimate the same number. Although AUC is quite popular among data scientists, its technical details fall outside the scope of this book. But since the pairing test is easier to understand and is an equivalent measure, that's the term I'll use in this chapter.

But the model's ability to get the pairing test right 91 percent of the time translates to low performance outside the research lab, where there's no contrived scenario presenting such pairings. In the real world, employing the model would require navigating a tricky trade-off. For example, you could tune the model to correctly identify two thirds of all gay individuals, but it would also wrongly identify others as gay, errors known as *false positives*. In fact, it would commit many such errors, predicting incorrectly more than half the time it predicted someone to be gay! And if you configured its settings so that it correctly identified even more than two thirds, the model would exhibit such errors even more often.

This is because one of the two categories is infrequent—in this case, gay individuals, which amount to about 7 percent of the general population of males (going by the stats the Stanford study cites). When one category is in the minority, that intrinsically makes it more challenging to predict.

Besides, accuracy isn't a helpful benchmark here in the first place. Accuracy only tells you how often the model is correct.

> **Accuracy**: The proportion of cases a predictive model predicts correctly, that is, how often the model is correct.

For this project, it would mean nothing to achieve a bedazzling accuracy of 93 percent: Just classify everyone as straight. By doing so, you're correct 93 percent of the time, even though you fail to correctly distinguish anyone in the minority, the 7 percent who are gay. To improve upon this and correctly identify at least some of the minority cases would require trade-offs: the introduction of false positives and, in general, a lower overall accuracy.

The Stanford model succeeded to a certain degree—it could predict better than guessing—but then the researchers misrepresented its performance on the pairing test by calling it "accuracy." Voila! Journalists and their readers believe the model can "tell" whether you're gay or straight.

Some things are too hard to reliably predict. "Gaydar" as a popular notion refers to an unattainable form of human clairvoyance. We shouldn't expect ML to attain supernatural abilities either.

Accuracy: A Word So Often Used Inaccurately

The Stanford study is a perfect example of a common misstep that I call the *accuracy fallacy*, which greatly exaggerates ML's performance across domains. It leads the public to falsely believe that the system can achieve an unrealistic, "crystal-ball" level of performance—specifically, that it can distinguish positive and negative cases and generally be correct for both positive and negative cases. That's just not feasible for many noteworthy prediction goals. Since many important behaviors—such as whether you'll click, buy, lie, or die—tend to occur more rarely, they're particularly difficult to predict. No model could "tell" such things with high reliability in general.

In some cases, researchers perpetrate a variation on the accuracy fallacy: They report the classification accuracy you would get if half the cases were positive. For example, Emory and Harvard universities reported on a model that predicts the onset of psychosis with "90 percent accuracy," as evaluated on data from a world where 50 percent of patients are eventually diagnosed with psychosis. There's a word for measuring accuracy in this way: *cheating*. Mathematically, this usually inflates the reported "accuracy" a bit less than the pairing test, but it's a similar maneuver and far overstates performance in much the same way.

The accuracy fallacy scheme is applied far and wide, in reports about predicting criminality, suicide, job resignations, bestselling books, deep fakes, tsunamis, and heart attacks. This list continues on to be breathtakingly long. For my extended coverage of the accuracy fallacy, see this chapter's notes at www.bizML.com.

Reviewing these projects spotlights both good and bad sides of the ML industry. The studies each misrepresent predictive performance, but, collectively, they illustrate ML's wide, cross-industry applicability. Many of them are otherwise legitimate projects, having generated a sound, potentially valuable model. The only problem is in how they misleadingly convey the predictive performance that the model achieves.

The accuracy fallacy contributes to AI hype. By conveying inflated performance levels and referring to the technology as *AI* rather than *ML*, researchers exploit—and simultaneously feed into—the public's fascination with awesome yet fictional powers.

The responsibility falls, first, on the researcher to communicate unambiguously and unmisleadingly to journalists and, second, on the journalists to make sure they and their readers understand the predictive proficiency they're reporting. But given current circumstances, we must all hone a certain vigilance: Be wary about claims of "high accuracy" in ML. If it sounds too good to be true, it probably is.

The Bad Rap of Imperfect Prediction

As we rightly take down overblown claims of fantastical prediction, others wrongly take down some of the most astute, albeit imperfect, forecasting. Clear-headed quants who forthrightly communicate the limits of their model nevertheless face a threat to their reputation by reactionary misinterpretation.

Take the famous political election quant Nate Silver. Despite the impressive track record of his political forecasts, pundits crucified him after Donald Trump defeated Hillary Clinton in 2016 for the US presidency. Silver's election forecast had put about 70 percent odds on it going the other way. As the *Harvard Gazette* put it, "Stunned political pundits blamed pollsters and forecasters, proclaiming 'the death of data.'"

But that's simply unfair and unjustified. "70 percent" does not mean Clinton will clearly win. And a 30 percent chance of Trump winning isn't a long shot at all. Something that happens 30 percent of the time is pretty common and normal. And that's what a probability is. It means that, in a situation just like this, it will happen 30 out of 100 times, or 3 out of 10 times. Those aren't long odds.

Unless you're posing as a soothsayer, relaying uncertainty isn't a sin—rather, it's often a virtue. Clinton's 70 percent probability was closer to a 50/50 toss-up than a 100 percent "sure thing." The take-away from a 70 percent forecast for Clinton isn't that she's pretty much a shoe-in. No, the take-away is, "I don't know." At the time, most other prominent forecasts put Clinton's chances much higher—between 92 and 99 percent. Those models exhibited overconfidence. Silver's model didn't strongly commit. It expressed, first and foremost, uncertainty. But, unfortunately for him, his model's probability was widely misinterpreted as a definitive prediction, as if he'd made an absolute call.

Forcing Your Hand

Being noncommittal is often prudent. Silver's best defense against public misapprehension might be to simply abstain when there's a fair amount of uncertainty and to only disclose his prediction when a candidate's probability is very high.

But for many deployments of ML, it has got to commit even when the prediction is uncertain. It must land on yes/no decisions—over and over again. Who should we market to? *Those predicted to buy.* Who should we approve for a credit card? *Those predicted to always pay their bill.* As we'll see later in the chapter on deployment, the mapping from prediction to action isn't always quite that simple, but, ultimately, predictions drive operational decisions, and the decision at hand is often binary.

This sounds like a recipe for disaster. We know that, when it makes committed predictions, ML generally doesn't achieve performance anywhere in the vicinity of perfection. We know that high accuracy isn't even the right objective. And we know the price Nate Silver paid.

Happily, the game ML plays is more fortuitous than the game Nate Silver plays, the game of singular, one-off election forecasts where each receives public scrutiny. ML repeatedly predicts, affecting a tremendous number of operational decisions and accumulating a track record along the way.

The Value of Imperfect Prediction

> All models are wrong but some are useful.
>
> —Famed statistics professor George Box

With the act of repeated, frequent prediction comes great news: For many applications of ML, getting a good number of predictions wrong is totally okay. So long as it predicts better than pure guesswork, that's often more than sufficient to improve large-scale operations and boost the bottom line. I call this the *Prediction Effect* (introduced in my previous book, *Predictive Analytics*):

> **The Prediction Effect**: *A little prediction goes a long way.* The law of large numbers is on our side; predicting better than guessing is generally more than sufficient—when applied across many predictions—to deliver value.

In deploying ML, we enjoy the security of a numbers game that's much more reliable than a single forecast, any one of which is so easy to get wrong. Will the economy go up or down next quarter? Will we hit our sales quota next month?

In its business deployment, ML drives many repeated operational decisions. By playing the odds over time, the overall performance pans out well as the model's predictions improve organizational efficiency. Your project isn't evaluated by any one case—you need not fret over individual missteps.

Okay, so a model's potentially good enough to be valuable, but exactly how good? How do we measure and report on predictive performance? Having blasted accuracy as impertinent and often misleading, what's the right metric?

Lift: A Meaningful Measure of Performance

To measure predictive performance in a meaningful way, you've got to differentiate between positive and negative cases. Accuracy doesn't do this. It simply reports, "How often is the model correct?"—across positive and negative cases alike. Since it doesn't differentiate, it generally fails to report on performance in a useful way. A bad model can look good. For example, if only 1 percent of the customers buy, a model that predicts "no" for every customer achieves 99 percent accuracy but fails to correctly predict any positive case—that is, any customer who is going to buy.

To remedy this, a common approach is to measure how often the model correctly identifies *positive* cases in particular. For targeted marketing, for example, this means we measure how often the campaign will contact the right customers—those who will buy. For most projects, *positive* signifies the less frequent class—which is typically the one that is more valuable to correctly identify—such as customers who will cancel, debtors who will default, medical images that signify the presence of a disease, or patients who will experience a heart attack.

More specifically, we want to know how much more often the model identifies positive cases in comparison to just guessing. There may not be many positive cases—say, only 1 percent of customers will buy in response to marketing. Can a model identify a "hot pocket" significantly richer in respondents? If the model picks out a group who buy 3 percent of the time, then it has improved over randomly selecting by a factor of three. This is known as a *lift* of three. That sounds like a good group to market to.

Wait a minute—that model is hardly better than a blindfolded monkey shooting darts. Its predictions aren't very confident. Among the customers it identifies for marketing contact, only 3 percent end up buying. It doesn't have high confidence regarding any one customer who will respond. It can never proclaim, "This customer will almost definitely buy."

But improving prediction just this small amount goes a long way. Predicting three times as well as guessing can *more* than triple profit. Let's say the model has made these positive predictions for 25 percent of customers—that is, the group that it has targeted for contact makes up a quarter of the overall population. This could multiply profit by more than five, as in the example covered in the following sidebar.

The Profit of Response Modeling

For one example scenario, here's some back-of-the-napkin arithmetic that shows how a lift of three translates to profit multiplying more than five times over.

Number of customers: 1,000,000

Cost per contact: $2

Profit per purchase: $220

Number of customers who purchase: 1 percent

Profit without a predictive model—mass marketing to all the customers:

Overall profit = revenue − cost

= ($220 × 10,000 responses) − ($2 × 1 million)

= $200,000

Profit of marketing to (only) 25 percent of the customers, with a lift of three—targeted with a predictive model:

Number of customers: 250,000

Cost per contact: $2

Profit per purchase: $220

Number of customers who purchase: 3 percent

Overall profit = revenue − cost

= ($220 × 7,500 responses) − ($2 × 250,000)

= $1,150,000

Given the clear benefit of achieving a certain lift, let's define it properly:

Lift: A multiplier: How many times more often positive cases appear within a group identified by a predictive model, in comparison with how often they appear in general.

Lift captures the multiplicative improvement to operations, the increased return for your efforts, how much more bang you get for your buck. With a lift of three, marketing to the targeted group will be three times as effective. Auditing the flagged group of transactions will find three times as much fraud. Examining the demarcated patients will find three times as many positive diagnoses. In this way, lift quantifies a model's contribution, even for models that aren't highly confident or anywhere in the vicinity of a magic crystal ball.

Lift reflects a limitation intrinsic to every model: It predicts less confidently as you target larger groups. The more cases it predicts as positive, the lower the lift will tend to be. Lift is always measured in relation to the group that's been targeted by the model. In the response modeling example, the model achieved a lift of three for one quarter of the overall population—the 25 percent predicted most likely to buy. If you went further and took one half, the 50 percent predicted most likely to buy, the lift would be lower since this group includes individuals who didn't make the top 25 percent.

Alternatively, the model could achieve a higher lift if it were used to identify a smaller, more select portion of the population. For example, a model might achieve a lift of 20 for the top 1 percent of the population. But there's a downside to targeting only such a small group: In exchange for more confident predictions, you end up with a much smaller pool of prospects—for marketing, this means a smaller number of potential buyers to market to.

Even a Little Lift Helps a Lot

Sometimes it makes sense to target a larger group despite that meaning a lower lift. Let's turn to credit scoring, where the model predicts whether a credit applicant will default on their loan. Suppose 10 percent of applicants will ultimately default and your model attains a lift of 1.7 at 50 percent. That is, it identifies a full half of the population of applicants who are 1.7 times as likely as average to default—that is, 70 percent more likely than average. In this case, the targeted half has a 17 percent likelihood of defaulting.

Seventeen percent might not seem like the worst risk in the world—but it's high in comparison to the other, less risky half, which defaults at a rate of only 3 percent. How'd we get to 3 percent? The arithmetic is simple: The two halves must average to an overall 10 percent default rate. Since that's the average of 17 and 3, we know that the less risky group must have a 3 percent default rate.

What a world of difference! Translating this into profit drives it home—the following sidebar shows that we can choose between winning or losing hundreds of millions of dollars.

The Profit of Credit Scoring

> Number of loan applicants: 1,000,000
>
> Average loss from a defaulting debtor: $5,000
>
> Average gain from a repaying debtor: $500

The model predicts half the applicants to be high-risk, with a 17 percent default rate, and the other half to be low-risk, with a 3 percent default rate.

If you approve high-risk applicants:

> Gain = 83% × 500,000 × $500 = $207.5M
>
> Loss = 17% × 500,000 × $5,000 = $425M
>
> Profit = gain − loss = −$217.5M (a loss)

If you approve low-risk applicants:

> Gain = 97% × 500,000 × $500 = $242.5M
>
> Loss = 3% × 500,000 × $5,000 = $75M
>
> Profit = gain − loss = $167.5M (a profit)

Lift applies universally, across ML applications. No matter what you're using ML for, lift serves as a fundamental, cross-disciplinary metric to report on a model's pure predictive performance. It provides a straightforward reading on how well the model can identify individuals relatively more likely to behave in a certain way.

But lift can only be calculated for a model once the model has targeted a group. This means it must commit to yes/no predictions that determine which individuals do and which don't belong in the group. Calculating lift—as well as profit—is possible only in relation to such a group. However, a raw model doesn't commit. For each individual, it outputs a probabilistic score that ranges across the spectrum of 0 to 100 percent. So, to determine who belongs in the targeted group, a threshold must be established in order to commit to a decision for each individual. Individuals with scores above the threshold belong in the targeted group.

So where do you draw that line?

An Illustrative Example: How Big Is Your TV?

When I teach lift in the classroom, I get everyone to stand up and be the data. Each person holds up a piece of paper with the size of the largest TV in their home and the group arranges itself into a row ordered by TV size:

People in order of their TV size (a zero means they have no TV). Those with a raised hand are subscribed to HBO.

Then I ask a question related to TV usage, such as, "Who has a subscription to HBO?" Not forgetting to thoroughly define the prediction goal, my question is actually more precise: "Who lives in a household that not only uses but pays for a subscription to HBO or Max?"

As you can see, the positive cases are more concentrated within the top portion—the left side—of this "human dataset." Let's calculate the lift for the top portion. Overall, 32 percent of the individuals have HBO (12 of the 38). But among the top ten, it's 70 percent. That's a lift of 2.2: That top portion has 2.2 times as many positive cases as the overall population.

Now, this classroom exercise oversimplifies in a couple ways. It's a comically small dataset, so the results are far from reliable. And the predictions are based on one and only one variable: TV size (that's what we call a *univariate* model). Plus, the scores aren't scaled to be probabilities, although they do serve to rank-order and calculate lift just the same. Despite these caveats, the visual effect and sample arithmetic illustrate the concepts for teaching purposes.

Why did I calculate lift for the top ten rather than, say, the top twenty? The answer is pragmatic. Perhaps I only have a marketing budget to contact that many. Or perhaps it's only worth marketing my product where the HBO subscription rate is at least 70 percent. Let's move on to some more realistic examples to explore further.

To Deploy ML Is to Triage and Prioritize

This story is the same for a large-scale modeling project: Cases more likely to be positive—and therefore more deserving of attention—rank more highly.

Instead of those thirty-eight people in a classroom, say we have a list of 100,000 customers. Here are the first four:

Name:	Model score:	Buy:
E. Siegel	85.628%	Yes
G. Clooney	85.626%	No
T. Mitchell	85.625%	Yes
T. Bayes	85.623%	Yes
. . .		

The model has scored each individual and we've ordered the list based on that score. The first one has a score of 85.628 percent. For response modeling, that would mean the model has calculated that as the probability they will buy if contacted—which would be quite a high probability, considering marketing campaigns tend to have very low response rates overall, in the single digits or even only a fraction of a percent.

Next, let's look at only the rightmost column, whether the case is positive or negative (yes or no). To show more, here are that column's values for the top 100 cases, showing the positive cases as a 1 and negative cases as a 0:

1011111111011110111100111111011111101111111111110111
1100111111101111110111101111100110111110110111

From a glance you can see just how dense with 1s this highly ranked portion is. The frequency of 1s doesn't necessarily decrease as you go through this list of 100, since this is only a small portion of a much longer list. But if we pulled out the 100 customers that are halfway down the list, we'd get a much different picture, with a more balanced mix of positive and negative cases:

1011010001011010111000100010111101110000110101110100
110110100100010111011010001011011101011011110000

And then the very final 100 would be mostly negative cases:

1100000000101001000010000000100000000010000100001000
0011000000010010000100001100010000100000100100

The same overall trend occurs across this long list as with the small set of students with TV sizes: Many positive cases appear early in the ranking, decreasing down to only small numbers at the end of the list.

A model achieves this same valuable effect across all kinds of ML applications. Its predictive scores serve to rank individuals so that the top portion is denser with positive cases and the bottom portion sees far fewer positive cases.

By ranking individuals, ML empowers the organization to triage and prioritize. Contact customers more likely to buy. Expend retention efforts on customers more likely to leave. Manually audit transactions more likely to be fraudulent. Inspect buildings more likely to catch fire.

This concept extends naturally to Internet search. It's the antidote to information overload. Google does you the favor of using ML to place an unwieldy number of items into a meaningful order. Facebook does too, ordering its Feed by predicting which of the many items your contacts have recently posted will be most of interest to you. Airbnb and Match.com follow suit, helping you sort through an oversupply of prospects, be they rental properties or romantic partners.

Of course, this also applies for the most literal of triage, medical triage. Tend first to patients scored by a model as more likely to decline in health or more likely to have a positive diagnosis. Reexamine patients predicted as more likely to be readmitted within the next year. Reach out to patients more likely to skip a prescribed medication or healthcare appointment.

In all cases, the model proactively targets according to risk or opportunity. It earmarks the individuals with the highest risk or potential gain—those worthy of investing limited time and resources.

Drawing a Profit Curve

The ranked list tells us where to draw the line—where to set the threshold—for driving a yes/no operational decision. Consider the decision as to whether to contact each customer with a marketing outreach. You can see the profit accumulated as we send a promotional brochure to customers, from most likely to buy down to least likely.

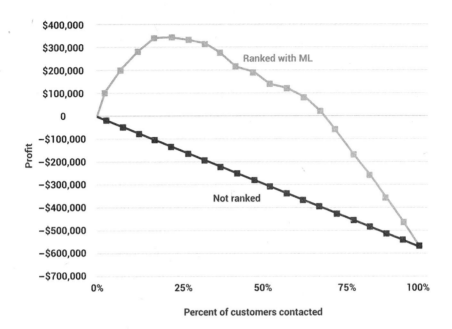

Percent of customers contacted

This is a typical *profit curve*. The horizontal axis corresponds with how far down the ordered list we've gone. As you proceed from left to right, you begin with those scored most highly by the model. At each position, the profit is calculated in the same way we did above in the sidebar on response modeling profit, based on how much we've spent to contact that many customers and how much we've gained from those who in turn responded with a purchase.

Following the upper curve, you can see the campaign's ups and downs. At the beginning, the more customers you contact, the more your profit goes up. Although you spend more to move along to the right—to contact more and more customers—you're getting enough positive responses to turn a profit. This is where you're getting the most bang for your marketing buck.

About one quarter of the way down the list, diminishing returns begin to set in. You've exhausted the most responsive portion of the list and you actually start to lose money—the cumulative profit begins to diminish as you contact more customers but you no longer elicit as many positive responses.

The overall marketing campaign is a bust if you actually contact the entire list. If you contact 100 percent of the customers, making your way to the far right of the graph, you end up with a loss of about $550,000.

For the example profit curve shown, if you have no model at all, you only lose. To visualize that situation, the straight, lower line shows what would happen without a model and therefore without any means to order the list. By following an effectively random order, you would lose money at a constant rate as you make your way through the list, so the lower line just keeps making a "beeline" down to the final end result. That line serves as a baseline for comparison. In contrast, the rise and eventual fall of the upper profit line is a testimony to how much value a model can deliver.

Either way, at the far right, you end up at the same place, losing about $550,000. This is because, if you're marketing to everyone, the order in which you do so doesn't matter—you always end up with the same overall loss, a negative profit. If you intend on just contacting everyone, you aren't targeting so there's no purpose to having a predictive model.

Turning back to the upper profit curve, you're probably feeling the urge to slam on the brakes, perhaps around the 25 percent mark. If you stopped there, your profit would be $350,000. That's often the best choice—but it's not an absolute. Sometimes, the marketing benefit of contacting more people takes a higher strategic priority, even if doing so isn't reflected in immediate-term profits. In that case, you may argue that stopping around 72 percent where you break even would be a much better choice than spending more than half a million dollars to contact everyone. That way, you basically get to market to almost three quarters of the list for free. Ultimately, the choice depends on the longer-term marketing strategy and other pragmatic factors at your organization. In any case, a profit curve like this one helps guide that choice.

The rise and fall of this marketing campaign seems to tell a tragic story. If only you knew the full story before playing it out, you could end the story earlier, quitting while you're still ahead—at the high point—instead of learning the hard way by winding up at the graph's deadly lower-right point. Is there a way to tell the story before living it?

Deploying Aggressively or Defensively—Your Cutoff
Makes the Difference

I'm going to blow your mind: The profit curve is only a projection. You can draw it before conducting the marketing campaign. Rather than tracking how marketing went, it estimates beforehand how it would go.

That's what data's for. It serves not only to train the model but to evaluate it and to plan its deployment. We'll dive into data in the next bizML step—"prepare the data"—but for now know this: Before deployment, models are provisionally tested on the same kind of historical data used to develop them.

Before deployment, you can draw this kind of projected curve just the same for most any ML project. To help decide how many individuals to target, you view the spectrum of options corresponding with how the model has ranked individuals. As you move along the spectrum, deciding how many to contact, approve for a loan, or audit for fraud, you often see the same pattern: an upward ride followed by a decline. There's a sweet spot, a Goldilocks zone, that's often the best place to stop. By establishing a threshold—a.k.a. a cutoff point—at that position, the model will then serve to be selective, targeting for treatment those who scored above the threshold.

In this way, one model provides a whole range of options. When you draw the line, you're establishing which option to go with in deployment—you're deciding precisely how to use the model.

For example, a response model could be applied either to increase revenue or to decrease costs. After all, increased efficiency can pay off in either way. To cut costs while maintaining current sales, target a smaller number of more highly ranked prospects. In comparison to mass marketing without a model, this could land the same sales even while spending less on the campaign. Or, to increase sales without increasing costs, spend the current marketing budget more wisely by targeting with the response model. In this way, you're contacting a more responsive pool. The difference between these two options comes down to where you draw the line, that is, how far you proceed down the list of ranked customers.

Similarly, fraud detection provides the same kind of trade-off options. For example, Citizens Bank developed a checking fraud model that could be used to either prevent 20 percent more loss or decrease their fraud prevention staff by 30 percent. On one hand, a set team of human auditors would capture more fraud if they spent their time on a more targeted pool of transactions that includes more fraud. This way, their time would be better spent. On the other hand, the model could maintain the current level of fraud detection with a smaller team of auditors. With a smaller team auditing a smaller group of transactions—but one even more densely filled with fraud—they'll catch the same amount of fraud as a larger team with no model. Once again, the choice between these options is enacted by where you set the threshold and draw the line.

You can think of these pairs of options as presenting a choice between a more aggressive or a more defensive deployment. The decision is a strategic one. Although there's some intense debate around which is best, there's no one-size-fits-all choice. If the economy is dipping or your company is having a tough quarter, you can expect a focus on cutting costs. And some ML applications are usually advertised more on the "defensive" side as cost-reducers, such as predictive maintenance and supply chain optimization. Other applications are more often sold as a means to increase revenue, such as price optimization and product recommendations. But, for any use case, there's always a range of options for model deployment.

In these examples, the bottom-line profit generated by a model has guided how to plan its deployment. We wouldn't use lift alone to decide where to draw these lines. So, why do we need a raw measure of predictive performance such as lift when a business metric always takes the lead?

Business Metrics versus Model Metrics

Business metrics such as profit or ROI drive ML projects. They define the business objective, inform the deployment plan, and measure project success.

In contrast, many business managers find reports of a model's pure predictive capability abstract and arcane. It feels irrelevant to them. Performance metrics like lift or the pairing test tell you the technical capability of a model and give you a reading of its analytical success—for example, lift tells you how much better it does than guessing—but they don't directly tell you the business value model deployment will deliver.

So why not work only with a directly relevant business metric like profit and drop technical metrics like lift entirely?

Measuring a model's predictive power may be abstract, but that's also an advantage. Model performance metrics apply universally, telling you how well your model predicts no matter the application area or industry. Given ML's wide applicability, such metrics establish a language that all data scientists—and their tools—can always speak, across projects, across industries. And they serve to validate and debug a raw model before involving the complexities of its intended business context. Let's put a name on this concept:

> **Predictive performance metric**: A measure of a model's pure predictive performance, such as lift, accuracy, or the pairing test. Such a metric serves to evaluate a model's *technical* performance but does not directly assess the business value of a model. Therefore, when possible, this raw metric is translated into a business metric.

Among the many performance metrics commonly in use, this chapter focuses on lift because it's a good place to start for many projects. As a "predictive multiplier," it's intuitive and, for many projects, translates to profit with straightforward arithmetic, as we've seen.

Beyond lift and the pairing test, predictive performance metrics also include *f-measure*, *precision*, *recall*, and *sum of squares*. I won't take space in this book to cover them, but know that they each follow the same trend: They're relatively arcane to the business leader and yet each has their time and place for the data scientist.

Despite the guidance that model performance metrics provide to the technical model development process, at the end of the day, business metrics rule. They have the final word. And they appear first within a project summary. For UPS's delivery-prediction project, I didn't make you wait until chapter 6 on deployment to learn of the business-metric gains. I gave it to you up front in this book's introduction: $35 million and 18.5 million driving miles saved per year.

Which brings me to the next point: Profit isn't the only business metric. Savings, revenue, return on investment, marketing response rate, and debtor default rate are just a few others. How do you choose?

Business Metrics: Key Performance Indicators

> The most important metric for your model's performance is the business metric that it is supposed to influence.
>
> Wafiq Syed, Data Product Manager, Walmart

When it comes to business metrics, we don't have to reinvent the wheel. Let's go "old school." Your ML project's business objective is to optimize something and that something is a KPI:

> **Key performance indicator (KPI)**: A measure of operational business performance that is key to a business's strategy. An ML project's business value is expressed as a KPI improvement. Aka *business metric*.

The word KPI may burn your ears with the sound of twentieth-century jargon. It's often associated with the older domain of *business intelligence*. After all, KPIs are simple, both conceptually and mathematically. They're usually just about counting something.

But KPIs matter the most because they report on the most fundamental notions of success at an organization. A KPI measures the degree to which a strategic objective has been achieved. As such, business leaders will feel more familiar and comfortable with a KPI than a predictive performance metric. And they'll be more excited about it.

The KPI will drive the ML project, so select one that:

1. aligns with strategic objectives;

2. compels stakeholders in order to achieve buy-in for the ML project; and

3. is measurable, in order to track ML success.

Although revenue and profit are often obvious choices, sometimes it's short-sighted to go so directly for the "gold." For example, what's most important when decreasing customer attrition? Retaining only the most valuable customers? Not necessarily. The absolute rate of defection across all customers—including those who barely turn a profit—speaks to the integrity of your business and the extent to which the customer experience satisfies.

Fraud detection brings up the same question. In the immediate term, it pays to intercept the biggest fraudulent transactions—so you could make it a priority to preempt perpetrators about to grab $50,000 worth of goods rather than only $50. But there is also value in punishing offenders large and small in order to decrease the overall crime rate and protect the fundamental integrity of commercial transactions.

How about often-heard, lofty business objectives like "meet today's escalating customer expectations" or "enhance user experience"? These alone aren't specific enough to fulfill a KPI's third requirement: *Be measurable*. They must be translated into well-defined quantitative measures such as "frequency of product returns" or "customer satisfaction conveyed on a survey."

We seem to have our left and right feet in two very different worlds. KPIs pursue business goals. Performance metrics like lift pursue pure predictive power. How do the two relate?

Distinguishing False Positives from False Negatives

The quantities that data scientists are trained to optimize, the metrics they use to gauge progress on their data science models, are fundamentally useless to and disconnected from business stakeholders without heavy translation.

—Katie Malone, *Harvard Data Science Review*

Sometimes, translating from a performance metric to a KPI is straightforward. Within a couple of sidebars earlier in this chapter, we've seen that applying some arithmetic based on the model's lift serves to calculate profit—for response modeling and for credit scoring. The intuition behind these calculations is straightforward: Lift tells us how many times better the model has predicted than guessing, and that multiplier is how much more often we get to seize on more profitable cases or avert more costly cases.

More generally, we often can span a mathematical bridge from technical performance to business performance by incorporating the price you pay when a model predicts wrongly. You incur a *misclassification cost* for two different kinds of prediction error.

> **False positive (FP):** When a predictive model says "positive" but is wrong. It's a negative case that's been wrongly flagged by the model as positive. Also known as a *false alarm* or a *false flag*.

> **False negative (FN):** When a predictive model says "negative" but is wrong. It's a positive case that's been wrongly flagged by the model as negative.

A false positive

A false negative

As we've seen, accuracy is a blunt instrument. It's one thing to know a model is wrong, say, 12 percent of the time. That's the same as saying it is correct 88 percent of the time; that is, it's 88 percent accurate. But it's another thing, a much more helpful thing, to separately break down how often it's wrong for positive cases and how often it's wrong for negative cases. Accuracy doesn't do that (nor does the pairing test).

That's what FNs and FPs are for. A FP is when the model says "positive" but is wrong. It's a negative case that's been wrongly flagged by the model as positive. This is also called a *false alarm*. The story about the boy who cried wolf is about him intentionally generating false positives. A FN is when the model wrongly says "negative." It has overlooked a positive case.

Calculating Savings Based on Misclassification Costs

So how do we assign a cost to each of these kinds of misclassification? That comes down to how much each kind of error matters. For almost all projects, it matters a different amount for a FP versus a FN.

Take fraud detection. When your bank's model wrongly blocks your legitimate credit card transaction as if it were fraudulent, you're inconvenienced. That's a FP. This could cost the bank $100 on average, given that you may turn to another card in your wallet—not only for the current purchase, but in general.

The other kind of error is worse. When the bank's model wrongly allows a fraudulent credit card charge to go through, that could cost the bank $500 on average, as the criminal gets away with the contraband. That's a FN.

These costs are no small deal. Global payment card fraud losses have surpassed $28 billion annually. The cardholder or an eagle-eyed auditor may notice the bogus charge later, but for card purchases, if it isn't caught by a model on the fly, it's in the wind. In the United States, the bank is usually liable for this loss.

By determining the two misclassification costs, we establish a cost-benefit analysis not only for the entire project but for each individual decision about whether to hold or authorize a transaction. Next, we'll add those individual costs up to calculate a KPI for the overall project: *cost savings*.

With no fraud detection model deployed, a medium-sized regional bank could be losing $50 million per year. The following sidebar steps through a scenario for such a bank, showing the savings achieved by introducing a fraud detection model.

The Cost Savings of Fraud Detection

Consider a bank that has issued 100,000 credit cards and each sees an average of 1,000 transactions per year, with one in 1,000 being fraudulent. To summarize:

> Annual transactions: 100 million
>
> Percentage that are fraudulent: 0.1 percent
>
> Annual fraudulent transactions: 100,000
>
> Cost per fraudulent transaction: $500 (the FN cost)
>
> Annual loss from fraud: 100,000 × $500 = $50 million

It looks like crime does pay after all. But before you quit your day job to join the ranks of fraudsters, let's see what fraud detection could do to improve the situation.

If the bank is willing to treat two of every 1,000 attempted transactions as potentially fraudulent—holding the transaction and possibly inconveniencing the customer—then the onus is on a fraud detection model to flag which transactions should be held.

Let's assume the model attains a lift of 300. That's a lot higher than the lift of, say, 3 that we discussed in a previous example. But remember that lift is always relative to the size of the targeted group. In this case, we care about the lift only among the very top, small sliver of transactions scored as most likely to be fraudulent—the top 0.2 percent that will be blocked. We won't block any attempted transactions other than those, so that sliver is all that counts. Given that it's such a small portion, a high lift is feasible—a model's scores can potentially sort transactions well enough so that at least the very top portion includes a high concentration of positive cases.

First, we need to calculate how many errors occur, broken into FPs and FNs—how often the model wrongly blocks a legitimate transaction and how often it lets a fraudulent transaction slip by. Here's the breakdown:

Transactions blocked: 200,000 (two per 1,000)

Percentage blocked that are fraud: 30 percent

(Lift × overall fraud rate = 300 × 0.1 percent)

Fraudulent transactions blocked: 60,000 (30 percent × 200,000)

FPs—legitimate transactions blocked: 140,000 (200,000 – 60,000)

FNs—fraudulent transactions allowed: 40,000 (100,000 – 60,000)

This model is often wrong, but extremely valuable. When it blocks a transaction, it's usually wrong—only 30 percent of the blocked transactions are fraud. This isn't unusual. Since fraud is so infrequent, it would be very difficult to correctly detect some cases without also incorrectly flagging legit transactions even more often. With legitimate transactions—that is, negative cases—so prevalent, even misclassifying a small portion of them means a lot of FPs.

So the best we can hope for from a model is that it provides an advantageous trade-off between FPs (less costly) and FNs (more costly). To calculate the bottom line, we add up the costs. We've already established the cost for individual errors:

Cost of a FP: $100 (inconvenience to a customer)

Cost of a FN: $500 (fraudster gets away with it)

So we need only multiply these costs by how often they're incurred:

Aggregate FP cost: $14 million (140,000 at $100 each)

Aggregate FN cost: $20 million (40,000 at $500 each)

Total cost with fraud detection: $34 million

We've cut fraud losses by $30 million (from $50 million to $20 million), but introduced $14 million in new costs resulting from FPs. Clearly, this is a worthy trade-off.

Overall cost savings: $16 million ($50 million – $34 million)

If you would like to access a spreadsheet with these calculations and try out different scenarios—such as varying the model lift, the number of transactions held, or the cost of each FP and FN—see the notes for this chapter at www.bizML.com.

Fraud detection only achieves the cost savings shown in the sidebar above by sacrificing a little accuracy. The model in that example is 99.8 percent accurate, slightly lower than the 99.9 percent accuracy of a "dumb" model that simply assumes every transaction is legitimate (and therefore takes no action to prevent fraud). In this case, a less accurate model was actually better.

To understand why, just revisit accuracy's fatal flaw: It doesn't distinguish between different kinds of errors, treating FPs and FNs as equally bad. Since it doesn't account for different misclassification costs, accuracy oversimplifies for all but very rare ML projects where the costs don't differ. For most projects, accuracy is a red herring.

Beyond delivering business value, fraud detection pursues a societal objective: It fights crime. In the example shown, it blocks more than half of the attempted fraudulent transactions. In so doing, it meets the expectations of consumers. Although citizens at large sometimes bristle at being predicted by models—electronically pigeonholed to receive bad ads or bad credit—when it comes to using their card, many consumers welcome prediction, gladly withstanding the occasional blocked transaction. Instead, they bristle when there's *no* predictive intervention and they're charged for a purchase they never made. Although they may not quite be cognizant of it, the typical cardholder has an expectation of fraud detection.

In the next chapter, we will return to this vital ML application, payment card fraud detection. But first we must finish our discussion of misclassification costs.

Subjective Costs: Misdiagnosis versus Missed Diagnosis

Establishing misclassification costs is critical. By doing so we can bridge a precarious gap, translating from pure predictive performance to a business KPI. Analytics consultant Tom Khabaza has been telling us how important that is for a long time—ever since we called ML "data mining." His *Value Law of Data Mining* states, "There is no technical measure of [a model's] value. . . . The only value is business value."

But sometimes it's next to impossible to set the costs. Take medical diagnosis. If you mistakenly tell a healthy patient they've just had a heart attack, that's really bad. You can imagine the unnecessary stress, as well as the unnecessary treatments possibly administered. But if you mistakenly fail to detect a real heart attack, that's worse. You let a serious condition go untreated. How much worse is a missed diagnosis in comparison to a positive misdiagnosis? A hundred times worse? Ten thousand times? Someone's got to put a number on that and I'm glad it's not me.

But we all must navigate this kind of judgment call more often than you may realize. For example, consider deciding whether to take a COVID-19 test during the pandemic. Like predictive models, the at-home antigen tests are imperfect. In the spring of 2021, during the delta variant surge, an at-home antigen test briefly convinced me that I had COVID, but a subsequent series of negative antigen as well as a few negative PCR tests showed me that it had almost definitely been a FP. That FP had a cost: Before realizing I was in the clear, my life was greatly disrupted as I quarantined from my family and canceled various plans.

The cost of inconvenience may have paled in comparison to the cost of COVID going undetected, but it wasn't a zero cost. Weighing these costs was rarely made explicit, but an implicit disagreement in how they compared underlay contentious public disputes around COVID policies and etiquette. In fact, when the United States cut the recommended quarantine period in half in December 2021, it was partly because people were intentionally avoiding testing in order to avoid the consequences of a potential FP.

The same challenge applies for predictive policing, where rearrest-predicting models inform sentencing as well as decisions for or against paroling an incarcerated convict. A FP means someone stays in prison longer, even though they will not offend again. A FN means someone is set free sooner, even though they will commit a crime again. The quandary of how to determine the relative costs of these two kinds of errors rests at the heart of justice. And when one race, ethnicity, or other protected group more often experiences injustice by way of a model—that is, when the model commits FPs more for one group than another—it's called *algorithmic bias*. For more on ML ethics as a general topic, I provide an overview at the end of this book's conclusion. For a deeper dive on algorithmic bias and other topics in ML ethics, see my articles and videos at www.civilrightsdata.com.

For many business applications of ML, we have it much easier. The misclassification costs are often self-evident, based on business realities such as the cost of marketing, the cost of fraud, or the opportunity cost for each missed customer who would have responded if contacted.

But not always. Even spam detection can go wrong and cost you something immeasurable, such as a missed job—or even a missed date with someone you would have ended up marrying. A FP means you may miss out on an important message, and a FN means you have to manually filter spam out of your inbox. There's no consensus on how to best determine the relative costs of these two, but whoever is behind your spam filter made that determination—if not explicitly, then by allowing the system to effectively default to something arbitrary.

Sometimes decision makers must quantify the unquantifiable. They must commit to specific costs for misclassification errors—despite the subjectivity and ethical dilemmas. Costs drive the development, evaluation, and use of the model. "Be sure to assign costs for FPs and FNs that are directionally better than just passively assuming the two costs are equal," industry leader Dean Abbott told me, "even when you don't have a truly objective basis for doing so."

Challenges Translating from Predictive Performance Metrics to KPIs

The analysis paralysis we may experience when setting costs is not the only obstacle. Sometimes, to translate from raw predictive performance to potential KPI improvements, you've got to make some audacious assumptions. Some projects require a presumptive leap in order to estimate the expected business value deployment will deliver.

I took such a leap when pitching deployment to my first client, gay .com. Although they didn't go for it, you can bet I pitched the churn model's potential value. At 27 percent, my model showed a lift of 1.5, which I translated into a potential profit of $286,000. That's where the beautiful profit curve that I showed them came to a peak, forecasting the effect of a targeted marketing campaign offering a discount to customers most at risk for cancellation.

But the value of churn modeling is harder to forecast than the value of response modeling. A response model is trained on data recording a previous marketing campaign, but a churn model is only trained on the history of who did and who didn't cancel. Gay.com hadn't conducted a campaign to retain customers, so even if the model promised to target likely defectors well, we could only guesstimate how many of those defectors would change their minds after receiving a discount meant to keep them around. For gay.com, I assumed the company would gain an average of $100 in value for every would-be defector offered a $25 discount. The profit calculation also had to take into account that the company would lose $25 in revenue for FPs—customers who were provided that discount but weren't actually going to cancel.

Although gay.com didn't deploy, my second client, EduPay, courageously launched my ad targeting models—despite an equally high dose of uncertainty. Remember that, for EduPay, I generated 291 different models, each predicting whether the user would respond to its corresponding ad. I could see how well each model predicted, but that only indirectly spoke to how much deployment would pan out financially. How much would using these predictions improve over the existing process of displaying ads that are popular across users in general? Until we conducted the ultimate experiment that is model deployment, we couldn't be certain how much value it would generate.

To Launch Is to Take a Leap

At UPS, this uncertainty only added to the pressure Jack Levis felt. It magnified the challenge presented by the Innovator's Paradox—that the more novel or radical an idea, the greater the struggle to gain support for it. On top of that, Jack now faced the *Deployment Paradox*:

The Deployment Paradox: For some projects, the business value
of improved decisioning is hard to estimate before deployment; it
can only be reliably established *after* you deploy.

Jack's model was technically sound, predicting up to 93 percent
of tomorrow's deliveries correctly. And he knew how to estimate the
misclassification costs. Each FP cost a good deal, on average, since the
resulting plan would include an unnecessary delivery destination. FNs
weren't as bad, since it was often relatively easy to later correct a truck's
plan by incorporating a new destination.

But Jack wanted to translate this to business metrics. As he put it,
"I wanted to know what more accuracy's worth, moneywise." However, for UPS's complex delivery-planning system, this translation
wasn't straightforward. Delivery prediction injected insight into UPS's
complex planning system. How much this more sophisticated method
would improve operations in business terms was something you could
only roughly approximate beforehand. The pertinent KPIs included
annual dollars and driving miles saved. They also included the more
wonky metric *stops-per-mile*, which increases as truck planning becomes
more efficient—the more densely a route is packed with deliveries, the
more value is generated from each mile of driving.

Jack made rough estimates, but he would only find out the ultimate
value of his optimization system by tracking its performance during
deployment. When we get to the topic of deployment in chapter 6,
we'll see the ups and downs his system experienced. And we'll also see
EduPay's deployment results in that chapter as well.

But we have a couple more steps to go before deployment. In the
next chapter, we prepare the data on which the model will be trained.

4 Fuel

Prepare the Data

Data trumps the algorithm. Machine learning algorithms may be the fun, sexy part—everyone wants to crash that party—but improving the data is where you usually get the greatest payoff. Data is the source of predictive power. It encodes the prior happenings of the world, the experience from which ML will learn. ML software is only as good as the data you give it.

To make use of the data you have, you've got to expertly reconfigure it into *training data*, which takes a simple form: Whatever you want to predict, you need a bunch of both positive and negative learning examples. But despite this simplicity, training data is no easier to prepare than Michelin-starred cuisine. Whatever form your existing data has come to take, it probably wasn't accrued with ML in mind. As a result, preparing the training data usually represents the ML project's greatest technical bottleneck.

So why is data prep so commonly underestimated, underplayed, and undervalued? How do business-side priorities drive data requirements? How much data do you need? And how do you know which learning examples are positive and which are negative—where do you get those labels? Finally, what kind of noise in the data kills ML and what kind is copacetic?

The BizML Practice:

1. *Value*: Establish the deployment goal.
2. *Target*: Establish the prediction goal.
3. *Performance*: Establish the evaluation metrics.
4. *Fuel*: Prepare the data.
5. *Algorithm*: Train the model.
6. *Launch*: Deploy the model.

Scott Zoldi fights crime across the globe. His superpower is data—and an unprecedented, innovative process to amass that data.

He's got his work cut out for him. Every day, hordes of criminals work to exploit systemic vulnerabilities in how you and I shop. Their relentless work chips away at the very integrity of consumer commerce at large.

I'm talking about fraud. Crooks obtain your card details so that they can perform a transaction and make off with the spoils. In 2021, payment card fraud losses reached $28.58 billion worldwide. The United States suffers more than any other country, accounting for more than a third of that loss. To make matters worse, fraud increased during the pandemic, in part due to the increase in "card-not-present" virtual transactions. Some called it the "scamdemic."

Scott is FICO's chief analytics officer. He oversees the world's largest-scope anti-fraud operation. Day in and day out, his product Falcon screens all of the transactions made with most of the world's credit and ATM cards—2.6 billion cards globally. With Falcon, banks and other financial institutions can instantly block suspicious purchases and withdrawals.

This capability hinges on machine learning—and it demands an impressive dataset. As we saw in the last chapter, a fraud-detection model must predict well, striking a tricky balance so that it recognizes a lot of fraud and yet does so without incurring too many false positives. To this end, the data must fulfill exacting requirements. If you visualize the data as a simple table, just a big spreadsheet, it must be *long, wide,* and *labeled*—here's what I mean:

1. *Long.* You need data about real transactions—a lot of them. This list of many, many example cases from which to learn must be a *long* one. And by including a broad assortment of cases from around the world, the data can be *representative*. Each case composes a row of the data.

2. *Wide*. You need revealing information about each case, including behavioral characteristics of both the cardholder and the merchant. These are the factors on which a model will base its predictions. Since each row enumerates all these factors, the data is also *wide*. Each factor composes a column of the data.

3. *Labeled*. ML software needs many known examples of fraud from which to learn, prior transactions that have been designated as such. How do these cases get labeled? The fraudsters who perpetrated these crimes know which are which, but they have not, so far, been cooperative. This means we need humans on our side to manually *label* many examples. These labels typically make up the rightmost column of the data.

Wide data has more information about each case

E-commerce	$125	Not-present	$250/day		Yes
Grocery	$17	Chip	$700/day		No
Clothing	$275	Swipe	$25/day		No
Pharmacy	$27	Tap	$150/day		Yes
Utility	$59	Not-present	$75/day		No
Airline	$782	Not-present	$35/day		Yes
Hotel	$1,221	Chip	$100/day		No
Restaurant	$76	Tap	$40/day		No
Pharmacy	$32	Swipe	$275/day		No
Grocery	$112	Tap	$400/day		No
E-commerce	$43	Not-present	$80/day		No
Restaurant	$82	Chip	$30/day		No
Utility	$26	Not-present	$100/day		No

Long data has many cases

Such a dataset sounds almost impossible to acquire. It could only be sourced from multiple banks across the globe. And even if you somehow convinced these institutions to cooperate and obtained a representative slew of example transactions, the fraudulent ones aren't going to label themselves.

To obtain this data, Scott's got to align the stars.

The Lifeblood of Optimization

Most people think data is boring. The word "data" is a deal-killer at cocktail parties. I know this from personal experience . . . I have the data.

But data isn't just an arcane bunch of 1s and 0s. It's a recording of history, a list of prior events. It encodes the collective experience of an organization from which it is possible to learn, analytically, how to predict.

Preparing the data represents both the most meaningful and the most mundane sides of an ML project. Although it isn't the "rocket science" part, it's how you construct predictive potential. It's the most time-consuming technical step, a discipline unto itself often requiring a specialist known as a *data engineer*. Data prep is typically estimated to demand around 80 percent of an ML project's technical efforts and generally takes longer than expected. It is habitually underestimated.

But it's worth it! The resulting training data is the fuel that powers ML. So even business leaders must become conceptually familiar with its consequential—and simple—format.

Data prep as a topic is strikingly neglected and generally untaught. Its omission is an understandable but costly mistake that afflicts ML as a field: Newcomers flock to the excitement of hands-on model-training with little thought as to how the data—and its requirements—were conjured in the first place.

Perhaps more surprisingly, educators and other leaders encourage rather than correct this misguided path. Most technical ML books and courses take data prep for granted—they skip right past it. On day one, your first step is to load the data into the ML software. The presumption is that the data is all set and good to go. But this presumption is false.

There is no Santa Claus; you must gather and assemble the training data yourself.

As Cisco's chief data evangelist Jennifer Redmon put it, "New data science graduates have a false sense of security that the data they'll receive will be sound."

By not skipping past the steps up to and including data preparation, we properly plan for deployment. In particular, step 2 established a prediction goal worthy of deployment—based on an informed, socialized, and ultimately greenlit project—and the training data, in turn, will embody that prediction goal; the way you pursue that goal is by preparing the data accordingly.

What Training Data Looks Like: Rows and Columns

Before we see how Scott and his team at FICO get their data, let's first cover the basics by way of a simpler, more typical story.

When I pulled together the data for the EduPay project, I had it a lot easier than Scott. The data I needed was readily at my fingertips. This is typical. After all, most ML projects seek to optimize only the business, not the whole world, so you don't have to search the world over for data. Internal data is enough—at least to get started.

Even so, data prep always presents a great challenge. Preparing the necessary *long*, *wide*, and *labeled* dataset from internal sources is already hard enough without pulling it together from across organizations.

Data serves the prediction goal so that goal must be reflected within the data. The prediction goal determines what the data consists of and affects its length, width, and labels. For EduPay, we had this goal:

> **Prediction Goal for Targeting Ads (EduPay):** Will the user respond to this ad if it is displayed?

I needed a long list of examples with both positive and negative cases, situations in which the user did respond to the ad and others in which they did not. Here are three sample rows of data pertaining to an ad for a university:

Already seen	Grade	Gender	Opt-in emails	US citizen	email	Num majors	SAT written score	Responded
Y	10	M	Y	Y	yahoo!	1	600	N
N	14	F	N	Y	gmail	0	520	Y
N	12	M	N	N	hotmail	2	710	N

Three rows of training data for modeling response to an ad for a university. Only a sample of the input variables (columns) are shown. Grade 14 means the second year of college.

Each row tells a little story from which the modeling process will learn. For example, in the first row, a user who had already previously seen the ad was in tenth grade, was male, had opted in to receive emails, etc. When shown the ad, the user did not respond—a negative example. The rightmost column of outcomes—responded or didn't respond—holds the prediction goal established back in step 2.

You now know the main requirements for *training data*:

> **Training data**: The data from which modeling learns—that is, the data from which ML generates a predictive model. If the training data includes labels that indicate whether each example is positive or negative, it is *supervised training data*, which is required for *supervised ML*, the kind of ML covered in this book.

For most business applications of ML, the training data is really that simple: a two-dimensional table with one row per example. That's why it's also affectionately known as a *BOFF*—a "big ol' flat file." You may have started from a database of many interconnected tables, but you've got to get the training data into this two-dimensional form before feeding it into ML software. The software needs it that way (technically speaking, some ML software only requires it be described that way, as a *database view*, rather than it being reformatted into a BOFF, but the conceptual effort is the same). Preparing the training data is a prerequisite for modeling.

The most notable exception is for models that handle large files such as images or sounds, most commonly *deep learning* models, a.k.a. *deep neural networks*. In that case, the raw data for each case doesn't elegantly fit into a single row. So, for example, when you apply deep learning to classify medical images, the training data isn't conceived of as a two-dimensional table. Each case consists of an image—itself two-dimensional—and the image's label, such as whether there is a positive medical diagnosis. Such non-tabular data is sometimes known as *unstructured data*, with tabular data known as *structured data*. And yet, even with unstructured data, the broader concept still applies: It's a long list of positive and negative examples, even though each item in the list isn't structured as a simple row of data.

The Data Dictates What the Model Does

For a standard modeling project, the columns you set up in the training data determine the model's functional purpose, its input and output. The *target* column, conventionally but not always the rightmost one, is what the model will try to predict—it corresponds to the model's output. This is where the prediction goal established back in step 2 comes into play. By filling that column with those values, you're dictating what the model will predict.

Setting up the data is setting up the modeling. More than any configuration you may adjust when operating a predictive modeling software tool, it's the training data you give to that tool that determines what your modeling project is set to do, what the model will predict for each individual. You don't somehow characterize or "describe" the prediction goal to the modeling software. Instead, the one and only way in which you set up your established prediction goal is here and now in step 4, by filling in that column of data.

The rest of the columns are what the model will try to predict with—it will take their values as inputs. By putting those columns of data in place, you are stipulating that they will be available to the resulting model. These inputs will constitute the model's eyes and ears, the only pieces of information it will have in order to make its prediction about any one individual. They're the fodder the model will chew on before generating a predictive score.

As all data scientists know, the training data's input columns are technically known as *independent variables* (a.k.a. *features*) and the output column is called the *dependent variable*. But in this book, I'll just call them input and output variables or the inputs and the output.

By placing the input variables next to the output variable on each row, we juxtapose what was known at one point in time alongside the outcome that was found out later, which we'd like to be able to predict. This juxtaposition is what enables the system to discover how things known at one point relate to—and therefore are predictive of—something that will happen later.

To be clear, in this data, both the inputs and the output are already known—no prediction required. Rather, each row is an example from which to learn. Once we ultimately have a model, we'll apply it to cases that look similar to those in the training data, except the output will be as yet unknown. After all, that's the whole point of the modeling process: When the resulting model is used in deployment, all it will have is the input variables. The outcome will be part of the *ultimate unknown* that we call the future. Putting odds on that eventuality is precisely what we're making the model for.

Big Training Data for Ad Targeting

Of course, real training data grows much wider and longer than the small sample of three rows and nine columns shown earlier. It's wider because more model inputs means more to predict with. For EduPay, I pulled together thirty-three inputs—making the training data thirty-four columns wide if you also count the output variable. This included other elements from the extensive profile EduPay collects from its users, such as career objectives, school clubs, and military experience.

It also included inputs I designed in the hope they would provide novel predictive value, such as:

- The ratio of SAT verbal and math scores. This reflects whether someone is relatively "more verbal"- or "more math"-oriented. For example, if a user got 700 on verbal and 580 on math, the value for this variable would be 1.2.

- College category, as determined by keywords in the institution's name, for example, Ivy League, state school, university, technical school, or community college.

When you think up new inputs like these, they're called *derived variables* and your effort to add them is called *feature engineering*. This is typically an ad hoc, manual process that taps your own creativity. It's a key opportunity during the project for business insights to contribute, complementing the degree to which the machine will automatically form combinations that are useful for the predictive goal at hand. Some of the variables that people come up with would be difficult if not impossible for ML to derive automatically, so manually designing new inputs is an important part of the data prep process. Collaborating with business stakeholders to generate ideas for new derived variables often contributes a valuable impact.

Some of the most fruitful derived variables come from thinking outside the box. For example, a senior analytics leader, Brandon Southern, engineered inputs to detect fraudulent accounts several years ago when working at eBay. His hypothesis was that, since a fraudster would typically create many accounts, they would automate, and, therefore, the time to set up a new account would be relatively short, as would the time to get up a first auction listing. By introducing these factors as model inputs, the fraud detection system was able to better detect bots and contributed to roughly $20 million in loss reduction across multiple fraud detection projects that he worked on.

Although he now works at Amazon, Brandon reflects on the success of his creative insights and the critical need for adept feature engineering. "By 2025 more than 465,000 petabytes of data will be collected on a daily basis across the globe," he says. "However, only a fraction of a percent of this data is considered to be useful for analysis and models.

In order to locate the most useful attributes, feature engineering is a vital skill."

On the other hand, take care not to load the training data with too many inputs, since they don't come for free. Usama Fayyad, famed as the world's first chief data officer—first at Yahoo! and subsequently at Barclays—emphasizes tying your data investment to business value. After all, any and all inputs you include in the training data signify a long-term commitment: You'll need to maintain them moving forward for as long as the deployed model remains in play. "Keep the data pull narrow so you can justify it . . . and scope the data for only one use case at a time," Fayyad advised during a keynote address at Machine Learning Week.

But let's move on to the really big part of this big data: its length. Using data from eight months of EduPay's operations, I had more than 50 million training cases (rows). The company had accrued this abundance of experience from which to learn because this website was popular. It had witnessed many little episodes where a user was shown an ad and then either did or did not respond. And this was only for its *interstitial* ads, where the user was shown a full-page ad to either accept or reject before moving on. We didn't take on the regular, smaller ads that were embedded within normal web pages during this project.

However, this mammoth load actually broke down into 291 smaller training datasets. Remember, I was producing 291 models, one per ad—each trained to predict the odds that the ad would generate a response if displayed for the user at hand. This meant about 170,000 cases per training dataset on average.

170,000 rows may sound like a lot—but one can't blithely assume it's enough.

How Big Is Big Enough?

How much data do you need? Actually, that's not the right question. What really matters is how many *positive cases* you have, because those are more rare. You need a healthy mix of both positive and negative examples. If you have enough positive cases, you're sure to also have enough negative cases.

For example, if an ad's response rate is 1 percent, then 1 percent of the training data will be positive cases and the rest negative. In that situation, negative cases occur ninety-nine times as often as positive. With 1,000 positive cases, we'd have 99,000 negative cases.

Okay, then how many of those less common positive cases do you need? Often, you'll have thousands or even up into six or seven digits' worth. But when positive cases are very rare or your overall dataset is small, that's when this question arises. Responses to ads are rare. Cases of fraud are even more so, amounting to somewhere around 0.1 percent of card transactions.

This question can only be answered vaguely. For many projects, a few thousand positive cases is enough. Even a few hundred may be enough to make a project viable—it's still potentially worth trying. A few dozen usually puts you more into "research project" territory. But there are no absolutes here. You never know how well modeling will work until you try it. There is no concrete theory that provides an absolute answer since the factors at play are just too complex. These include how well your inputs (columns) serve the prediction goal, how well the model must perform to deliver value—depending on the business context—and just how "difficult" the prediction goal is. Some things are easier to predict than others. For example, asking the model to predict further into the future is usually harder than predicting the more immediate future, in that your model's performance would likely be lower for longer-term prediction.

Whatever amount of data may suffice to train the model, you actually need a bit more than that for the purpose of evaluating the model—you need *test data*. This is a sample of, say, 20 percent of the training data that's held aside during model training. Since it's not available during the model's formation, the test data serves as a basis to assess how well the model performs in general—in new situations that extend beyond the training cases used to develop it.

So, despite the mythology about drowning in "too much" data, the real scientific challenge comes when you have "small data"—especially when the number of positive cases is very small. For example, in health-care, certain diseases are very rare and applicable patient records can be hard to come by. Modeling on very small datasets is a rich research area. Studies have been published with impressively small counts of positive cases in the training data, yet other such modeling efforts just plain fail.

But for business applications, you often do have enough data—and this is not only because of today's general "data explosion." The reason is more specific: Large-scale operations that are worth the effort to improve are the ones repeated frequently and therefore are the ones for which you have accumulated a lot of data. It's an almost self-evident piece of good news: Anything you do a lot generates the data you need to improve that very process. The data you need has already grown organically in the course of conducting operations. For example, if you regularly conduct large direct mail marketing campaigns, you already have the history for such campaigns, including who was con-tacted and whether or not they responded. If you issue credit cards, you've tracked which customers turned out to be reliable debtors and which did not. Many times, ads have been displayed and the response recorded. Likewise, many transactions have been approved and the fraudulent ones subsequently spotted by unhappy credit card hold-ers. All the main things we do are worth improving and are the very things for which we've built up experience—that is, data—from which to learn.

Data abundance is a good thing; the more the merrier. So long as the data is representative of the kinds of cases that will be given to the model in its deployment, then more rows of training data, both posi-tive and negative examples, will only further help train the model.

But even as we gather as much data as possible, the ratio of posi-tive to negative cases is usually skewed. The data simply reflects reality: Those all-important positive outcomes occur less often. Will the lopsid-edness of your data throw things off?

Are Positive Cases Underrepresented?

Positive cases are usually more important—the reward for predicting them correctly is higher, as is the cost for predicting them incorrectly (*false negatives*—when the model has mispredicted a positive case as negative). The biggest wins come from identifying those rare customers who will respond to an ad and those rare transactions that are fraudulent.

And yet it's often negative cases that dominate the training data. Will this mean that model training prioritizes negative cases, generating a model that predicts better for them than for positive cases?

The answer is no—this problem has been solved. For some modeling methods, math can solve it by applying certain adjustments to the training algorithm that account for the imbalance. In other cases, it's best to reduce the abundance of negative cases, by essentially "throwing away" some of them—especially if there's so much data that doing so saves a lot of computing time in exchange for only a negligible loss in model performance. But there's a critical, yet often-overlooked caveat: If you do pare down the training data in this way, do not do so to the dataset used for testing the model after it's trained. That *test data* must retain the true, original balance, as it occurs "in the wild." If not, you begin treading on the *accuracy fallacy* territory we explored in the previous chapter.

In fact, having fewer positive cases often aligns with having a more valuable ML project. When you and your colleagues define the prediction goal—that is, what would be valuable for a model to predict, what it means to be a *positive* case—it's only natural that it occurs somewhat rarely. For example, if the goal is to predict which customers will cancel their subscription within three months, you could end up with, say, 15 percent positive cases. But if your prediction goal is those who will cancel within five years, that's a more common occurrence—it could be half of your current customer base. Knowing that a customer will cancel sometime in the next five years is less actionable. It doesn't clearly mandate any immediate response. Instead, knowing that a customer is likely to cancel in the near term more clearly suggests immediate action: Invest in retention activities to avert an imminent cancellation.

As another example, if you're targeting marketing, you may want to predict highly valuable respondents. That is, instead of just predicting who will make any sort of purchase, predict rare, special cases, such as who will buy a lot or buy higher-margin products. Likewise, for fraud detection, if you predict not any and all fraud but specifically more costly cases of fraud, you'll similarly be focusing the model on identifying the fewer, more important cases.

Not only do rare positives tend to be more valuable—there's also a cosmetic perk: Your model will look more impressive, since it will exhibit a higher *lift*. If an ad's response rate is 1 percent, the model could realistically flag a small group of customers with a lift of 10. That impressive lift of 10 means that, among that top portion of users most likely to respond, they do so ten times more than average—that is, 10 percent of the time. Last chapter, we calculated the value of a fraud detection model with a strikingly high lift of 300. That was only feasible given the very low fraud rate of 0.1 percent.

But if you're predicting something that happens half the time— for example, which customer will cancel within five years—you could never get a lift of 10. That would mean ten times the average cancellation rate: 500 percent. The maximum lift you could ever hope to report would be 2 and the model's lift will generally be somewhere below that maximum, so you might expect a lift around 1.2 or 1.3.

Generally speaking, reporting a low lift isn't as good a look. To seasoned data scientists, it isn't sexy. As it turns out, there's good reason—for many projects, this isn't only a cosmetic matter. A high lift aligns with high value. After all, finding positive cases, say, ten times more often than random guessing, can indeed be ten times more valuable to your business, since operations multiply their effectiveness that many times over: Auditors find ten times as much fraud in the pool of transactions they process; an ad gains ten times as many responses when displayed; and marketing hits ten times as many respondents.

So if this step, data prep, reveals that your positive cases occur close to half of the time, it may be a signal to circle back to step 2 and revisit the prediction goal, ensuring you've established as valuable a goal as you could.

But there's a balance to be struck. If you find that positive cases are too rare, leaving you with too small a count in the training data, you must backtrack to step 2 just the same—or somehow get more data.

It's about Time: Input Variables

Time is the wisest counselor of all.

—Pericles

Preparing the training data is a real ordeal because time matters. You're setting the stage to train a model, which later, when put to work, will input what is known at the time and output the odds on what's to come. This means the values you place in the input columns must reflect what was known at an earlier point in time than when the output column became known.

For example, let's say you pull from company logs that a user who's now in tenth grade responded to an ad one year ago. That serves as a learning case, but you've got to roll back the clock: This row of training data should show the grade as ninth, since that's what grade they were in when they responded to the ad.

It's a similar story when targeting marketing. Behavioral variables meant to serve as model inputs need the same kind of TLC. Consider a row of training data corresponding to a customer who was sent a marketing brochure last October 18 and at the time we knew he was male, lived in California, and had made ten purchases so far, as well as other factors. By two months later, on December 18, we knew that the outcome was positive—he had purchased the product.

October 18   December 18

Male, CA, 10 purchases...   Response: Yes

In this row of data, the input variables encode what we knew back when we made the marketing treatment decision, the point at which it could have helped to make a prediction. The final piece of information in that row of data, the outcome, was found out later—it's now also already in the past, but it came at a later point in time.

The potential for error comes in the fact that a lot may have happened since. Let's say it is now the following May. If records currently show they've made a total of fifteen purchases, we better have a method to calculate that, back in October, they'd only made ten purchases so far and include that amount within this row of training data.

Many time-related challenges like this arise and they're really easy to mess up. The mistakes are often called "leaks from the future," where an input accidentally relays information that could only be known later. For example, consider, when applying churn modeling, an input that indicates whether the customer has recently been included in a marketing campaign that was applied only to customers who hadn't canceled their subscription. In this case, the modeling method will very quickly "figure out" that this is a helpful input. The model's performance will appear high—but it's cheating. In deployment, the model wouldn't have access to that kind of "sneak peek" into the future.

As another example, one insurance company generated a model that seemed to be doing a great job predicting which policyholders would submit a high claim by incorporating the following pattern discovered within the data: Policyholders with an email address on file would be more likely to submit high claims. But it turned out that email addresses "leaked" the future because they had only been solicited from those who ultimately filed a claim. As a result, customers with an email address—that is, customers who had by now submitted at least one claim rather than none—were indeed more likely to have submitted a high claim. Once again, in deployment, the model wouldn't yet have access to this indicator of a claim. The discovery was circular, not predictive.

This "time leak" pitfall compromises the integrity of predictive modeling. With the relative future always readily available during data prep, it's all too easy to inadvertently provide that future to the very model trying to predict it. Such temporal leaks are a common "gotcha." Fortunately, they can often be noticed when the data scientist realizes that the model's performance just looks too good to be true.

Most People Aren't Prepared for Data Preparation

With these stringent "temporal" requirements on each input—that the value provided to the model reflects only what was known at a certain point in the past—data prep presents a tricky challenge. The data sources that you're repurposing weren't originally collected with ML in mind. In whatever form your data currently sits—distributed across tables, databases, and even systems—you must engineer its transformation into the form and format of training data. This transformation involves more than only achieving the requisite tabular structure. More than just the arrangement of the data, you're dealing with its very meaning.

Given these nuanced requirements, you can't fully automate data prep for a new ML project. Each project necessitates a specialized database programming task customized for the existing data sources in accordance with how they originated. Some analytics software may help automate limited portions of this task—and some analytics vendors will promise the moon—but no software can fully handle the meat of the matter for you. Once you've worked through how to program it, the programs may then automate data prep for future iterations to update the model. But for each project's first pass, you've just got to dig in and take this on. There's no running.

In some cases, you can employ a little trick to sidestep some of this complexity: *snapshotting.* This is the process of periodically capturing a "freeze frame" of the input variables, logging them as they look right now, so that later, after you've tracked the outcome that will serve as the output variable, you can then just tack this outcome onto each row of training data. In that case, you don't need to recreate the past values of the input variables—you've saved them as they were at the right time.

However, that trick doesn't always apply. If snapshotting hasn't already been in place for a while, you'll need to do it the hard way in order to make use of the data you currently have.

Common wisdom states that data prep takes a whopping 80 percent of the project's hands-on time, although some estimates are lower. A survey by the ML software company Anaconda said that data scientists spend 39 percent of their time on data prep—which is still more than the time spent on training and deployment combined, according to the same survey. But, for new ML initiatives, data prep is likely to take longer than that. On the other hand, if you're repeating a mostly established data prep process in order to update an existing model, the logical hurdles and troubleshooting probably won't pile on nearly as badly.

Understandably, the ML industry downplays all this. It prefers to portray itself as leading a glamorous life consumed with learning from data—rather than the grind of hacking that data together beforehand. Consultants, vendors, and proponents often leave this nuisance unmentioned in favor of discussing the sexy part, the modeling.

Beyond the "let's jump straight into the modeling" culture adopted by most of the popular courses and how-to books, one other major trend also inadvertently contributes to the costly devaluing of data prep: public ML competitions. The host of such competitions—most commonly a company called Kaggle—provides the training data and pits data scientists against one another to train the best possible model. This advances the state of the art and often provides sponsors with the best *crowdsourced* model that money can buy, but as a side effect, it perpetuates the misleading narrative that ML projects are all about the modeling. If you win a modeling competition, that doesn't necessarily mean you're prepared to lead an ML project to deployment.

To address this, the education tech firm DeepLearning.AI has launched a new kind of *data-centric* competition. "In most ML competitions, you are asked to build a high-performance model given a fixed dataset," proclaims this unique contest. "The Data-Centric AI Competition inverts the traditional format and instead asks you to improve a dataset given a fixed model." Competitors don't get to make the model—they only get to improve the data that's then used for model training. This could help course-correct a culture that's overly fixated on modeling.

Some Noise Kills ML, but Some Is Copacetic

Given its stringent demands, how disastrous is noise in the training data? In light of the credo "garbage in, garbage out," one might presume that any noise would kill ML—especially given the intricacy and delicacy of model training. If an ML algorithm is misled by bad data and develops a faulty model, the consequences could be dire—the stakes are high when you deploy models for fraud detection, financial credit scoring, and so many other pivotal application areas.

Well, you may be surprised. ML is actually quite robust in the face of certain kinds of noise, so there's only so much cleanup you actually need to do.

First, let's de-noise the word *noise*. It can mean two very different things. If data has errors—plainly incorrect values—that's one kind of noise. This is sometimes called *corrupt data*.

On the other hand, if the data shows values that appear random— since we have no way to predict them, no basis for understanding their origin—that's another kind of noise that doesn't necessarily involve incorrect values. It just reflects our human lack of knowledge. It's uncertainty. This is the kind of noise Nate Silver references in the title of his famous book, *The Signal and the Noise.*

So, when data appears senseless or random, that doesn't necessarily mean it's faulty. It just means we don't understand all the factors that have affected it. ML helps us understand the world a bit better by discovering trends in the data. It finds some signal, but it doesn't by any means eliminate all the noise.

On the other hand, noise in the sense of outright errors can be a concern. It's pervasive, stemming from many systematic issues. Maybe somebody mislabeled a field within one of the databases you merged in. Maybe the age for a group of customers was calculated based on the year of birth without looking at the month and day, so it's sometimes off by one, depending on the current month and day. Maybe missing (unknown) values were changed, incorrectly, to the value 0 when data was ported between systems that represent missing values in different ways. Maybe your sensors record only imperfectly. Or perhaps there's even been malicious, intentional corruption of values in the data.

Despite all this potential noise, things aren't so bad. Here's your saving grace: Noise among the output variable labels is detrimental to ML, but ML can generally withstand noise in an input variable, so long as the amount of noise remains consistent.

First and foremost, the output variable labels must be sound. They encode the prediction goal, which gives model training its direction. For example, if many customers who didn't buy are labeled as having bought, or vice versa, this is very much going to mislead the modeling process. The output variable's integrity is critical—it must align with *ground truth.* Not only does it guide the learning process; it also serves to evaluate model performance after the modeling is complete.

But on the other hand, ML is robust to noise within input variables. Basically, you can throw a lot of junk in there, and although it won't help model training, it usually won't hurt nearly as much as you might imagine. The reason is that the less helpful an input is, the less the model will rely on it. Taking that to the extreme case, if an input has so many incorrect values that it's effectively random junk, then a good learning algorithm will completely avoid using that variable—or at least lower its involvement within the model down to nearly zero.

To put it another way, incorrect input values just make for more noise. Inputs are already noisy in the sense of uncertainty. For the ML algorithm, noise is noise and it doesn't matter which kind of noise it is.

Picture it this way: If you're doing image classification, you can see that even if there's a fair amount of noise in the image, it is still easy for you to tell that it's a photograph of a person wearing a hat. A model can handle noise much the same.

A photograph of a person wearing a hat is still easy to discern with noise added.

It's a relief the world won't end if you have bad values among the input variables. But there's one major caveat: The level of noise must remain roughly consistent between the training data and the data input to the model during deployment. If incorrect values are provided when actually using a model, but a similar prevalence of errors wasn't present in the training data when developing the model, the model's performance won't hold up. So its input better not be wrong much more often than it was within the training data that the machine learned from in the first place.

This is just one part of a more general requirement: The training data must be *representative* of the data encountered during deployment. That is, the experience over which the machine learns must come from the same "world" or "universe" within which the model will subsequently be used. The learning cases must *represent* the same "reality." For example, if the model will target marketing for all customers in the United States but the training data only includes customers in California, that data won't be representative; you can't expect model performance to carry over to the rest of the country.

For the EduPay project, my data failed to meet this requirement. It wasn't entirely representative. For each ad, the training cases were a result of the legacy system's peculiarities. For example, an ad hoc method estimated how popular each new ad would be when first launched, which determined how often and on whom it was tested. I was working with the available data in hand, not conducting experiments to collect an evenly distributed battery of tests for each ad. Although there are technical approaches that might have helped, I didn't take any measures to address this imperfection in the data. In the chapter on deployment, you'll see how things panned out nonetheless.

Now that we've covered the various requirements for training data, let's return to FICO to see how the company pulls the data together for its global fraud-detection system.

FICO Cultivates Data without Borders

Scott Zoldi has a PhD in theoretical physics from Duke University. And he's formed a team of seventy more people with PhDs. Together, they generate the world's de facto system for detecting fraudulent card transactions. You, me, and most people with payment cards are relying on them.

Scott's antifraud operation isn't what FICO is most widely known for. Along with another one of his teams, Scott also oversees this country's *most famous* deployed model: the FICO Credit Score. Your FICO

Score determines your power to borrow. It's the most widely used credit score in the United States, employed by the vast majority of banks and credit grantors. It's a household name, and many understandably feel that their FICO Score is a central part of their identity as a consumer.

But FICO's fraud detection, which is normally invisible to us as consumers, affects us much more often. Named Falcon, this product is the biggest part of FICO's software business and affects most of us almost every day—every time you use your card. FICO evaluates financial power by day—and fights financial crime by night.

To meet this responsibility, it's important that the Falcon team gets the data it needs—some long, wide, and labeled data. To do so, it collects data from across a global network of banks.

This reliance on inter-enterprise data—collected from multiple companies—is atypical. Ordinarily, an ML project serves only the enterprise running the project. For such a project, internal data suffices, since the company has been tracking the very operations that the project aims to improve. In contrast, FICO isn't a bank. It doesn't process card transactions. Rather, it holds a rare, globally central, entrusted role across banks.

In 1992, Falcon was born of a radical move by a small group of banks: They decided to cooperate rather than only compete. At the time, a tremendous portion of all credit card transactions—almost 1 percent—were fraudulent. The fraud rate was only growing and threatened the entire industry. This looming crisis convinced financial institutions to overcome their raw capitalistic instincts and follow a call to arms for the universal good: to collaborate to fight crime, improve transaction integrity, and cut losses. Led by a company called HNC Software, they joined their data together, thereby multiplying their power to train fraud-detection models. Ten years later, FICO acquired HNC Software—and both Falcon and Scott Zoldi along with it.

Since then, Falcon's consortium has grown to more than 9,000 banks globally, all continually sending in anonymized card transaction details. FICO receives about 20 billion records—amounting to terabytes of raw data—each month, a petabyte every five years.

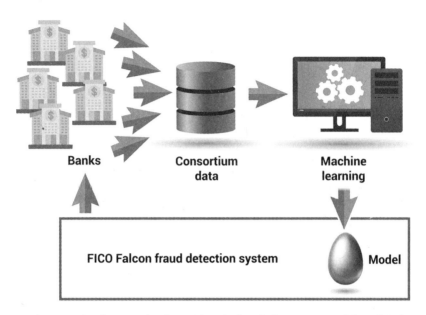

Banks   **Consortium**   **Machine**
                         data   **learning**

FICO Falcon fraud detection system   **Model**

Banks provide data to develop Falcon's fraud detection model and Falcon
deploys that model for each bank.

Banks can't benefit from Falcon without contributing to it. To be
a FICO customer that uses Falcon, you must also join the consortium
and share your data. Falcon has become so standard that, despite its
cooperative nature, it's a competitive necessity. To hold their position
in the payment card market, banks need Falcon's best-in-class fraud
detection—which they can access only by cooperating. In the end, this
levels the playing field: Even the smallest bank can deploy the very best
fraud-detection model.

Engineering Better Inputs for Falcon

For each card transaction, Falcon's model needs an informative but
reasonably succinct summary of all the "evidence" that could reveal
whether it's legitimate or fraudulent. That list of factors, each a potential model input, makes for a row of training data. It includes a rundown
of the cardholder's previous transactions, such as the total number of

transactions, number of cash withdrawals, and average transaction amount. These counts are broken down by different time ranges, such as for transactions conducted over the last day, week, month, and three-month period. The counts are also broken down by different types of merchants, such as restaurants and clothing stores.

But that's only the basic groundwork. Scott's team has made a fine art of feature engineering, manually designing more complex inputs that further help identify fraud. These inputs encode whether recent transactions mark changes in cardholder behavior, such as an extreme acceleration in spending or purchases from a completely new kind of merchant—for example, purchasing from a golf store when you've never done so before. At the same time, long-term tracking for each cardholder must recognize annual trends so they don't trigger false flags, such as when a family vacations every year in Florida.

Actually, it's a bit more "meta" than that. Even the most dramatic change in behavior doesn't necessarily reveal fraud since some card-holders are prone to change. They're inclined to do something unprecedented. For this kind of customer, purchasing from a golf store for the first time should not be interpreted as potential fraud. For them, change is nothing new. Some anomalies are benign. With the right inputs, a well-trained model will be able to discern *which sort of change in behavior corresponds with fraud.*

On top of all this, other inputs track updates to the cardholder's profile. For example, after you change your email address, it may turn out that you're a little more likely to conduct a transaction in a new-to-you country. With inputs designed to reflect this kind of change, the model can potentially learn when to be forgiving of such first-time behaviors. Remember, the aim isn't only to identify fraud but also to tame false positives by allowing legitimate transactions even when they're an out-lier for that cardholder.

Finally, other inputs help the model scrutinize aspects of the mer-chant. After all, it's not only the cardholder's history that reveals that an attempted transaction may be fraudulent. The goings-on at the merchant—across all cards—also provide clues. This includes whether the merchant has experienced a rise in attempted fraud or other unusual changes in how transactions are trending there.

It's Not Over Yet: Labeling the Data

Beyond well-engineered inputs, Falcon's training data needs one more ingredient: labels for the output. Each example transaction that makes for a row of data is incomplete until designated as either fraudulent or not fraudulent. Those labels will guide model training to do its job: *Generate a model that can discern positive cases from negative cases.*

Only humans can provide the labels. As I briefly touched on a couple of chapters ago, for detection, we don't get to benefit from "time will tell," as we do when predicting a future event. Time has told whether a user responded when shown a certain ad or whether a debtor has defaulted. In those cases, we get the label "for free." But for detecting a qualitative attribute for each case—such as whether it is fraudulent— each training example's label can only be determined by a person.

Manual labeling is labor-intensive and expensive. The expense especially racks up when it requires subject matter experts, such as doctors for establishing whether each example indicates a certain medical diagnosis.

On the other hand, problems that don't require special expertise, such as labeling traffic lights within images for an autonomous-driving project, can be outsourced on "crowd labor" platforms like Amazon Mechanical Turk for as little as a penny per case. But there's a dark side: Their largely unregulated working conditions "offer a bleak glimpse of what could come for a growing digital underclass," according to *Vocativ*. *Marketplace* calls this "the new factory floor of the digital age."

To make matters worse, fraud detection requires an immense number of labeled transactions because positive ones are rare. If the fraud rate is 0.1 percent and you want the data to include at least 10,000 positive cases, then you need to label 10 million cases as to whether each is positive or negative.

Don't fret! Falcon's training data manages to sidestep this costly bottleneck by relying on what consumers do naturally. With card fraud, if the consumer sees an erroneous charge, they complain. We cardholders and our banks are in effect already doing all the grunt work to label many cases of fraud in the course of just living our lives.

A key reason this approach works is that, with card fraud, banks can afford to learn the hard way. Since the detection system is imperfect, it allows some fraudulent transactions to go through. This generates a positive training case if the cardholder later complains about the unauthorized charge—even though it's then typically too late to prevent the fraudster's crime. The cost is absorbed by the bank, but the overall cycle is economically copacetic. No humans were substantially harmed in the process of this data creation.

In other domains, you can't do it that way. The missed, uncaught cases—false negatives—aren't nearly as allowable for an autonomous vehicle that would drive through a red light or a medical system that would miss a diagnosis. In those domains, you often can't avoid the need for additional manual work labeling many examples.

This "organic" labeling process for fraud detection, wherein people are essentially "following the money," prioritizes bigger cases of fraud over smaller cases. FICO treats only *adjudicated* fraud as positive cases, where the cardholder has formally certified that the transaction was fraudulent (whether it was them or the bank who'd noticed it in the first place). This means that suspected cases that never get adjudicated aren't labeled as positive in the training data, even if the bank had to write off the charge. Since folks tend to bother with adjudication more for larger-value cases of fraud, lower-cost fraud is less often correctly labeled and is therefore effectively deprioritized by Falcon's model. And that's tolerable since the *false negative cost* is lower for them.

On top of this manual labeling, many other positive cases are passively labeled: those Falcon has automatically spotted. A bank using Falcon blocks an attempted fraudulent transaction and the cardholder might never even hear about it. This is almost a circular process, since that positive example will then serve to train an updated model for Falcon, which identified the positive case in the first place. However, once again, natural cardholder reactions help correct the data. If Falcon was wrong—if it is a false positive—then the cardholder, whose legitimate attempt to transact was blocked, will often take action to get it approved and the case will wind up as negative in the training data. In that way, what the model got wrong will serve to improve the next version of the model.

Altogether, this provides plenty of positive examples for Scott's team. The number of labeled cases of fraud that they end up with approaches one million.

FICO Falcon Fights Fraud Fantastically

Falcon works. I consider it one of the world's most successful and widely impactful commercial deployments of ML. It screens all the transactions for 2.6 billion payment cards worldwide. That's two thirds of the world's cards, including about 90 percent of those in the United States and the United Kingdom. Seventeen of the top 20 international credit card issuers, all of the United States' 100 largest credit-card issuers, and 95 of the United States' top 100 financial institutions use Falcon.

Since its introduction, Falcon has reduced card fraud losses by more than 70 percent in the United States. With the United States currently suffering around $10 billion in annual fraud losses, that reduction is saving us in the vicinity of $20 billion per year.

Just as Falcon's data collection is distributed across banks, so too is its deployment. What's more, that deployment has to happen in real time, generating each predictive score in a matter of only milliseconds. These considerations are coming in chapter 6 on deployment.

But first, we turn to step 5, using our hard-earned data to train the model itself. How do we mere mortals design an algorithm that generates a predictive model from training data? That's one hell of a computer programming homework assignment.

5 Algorithm

Train the Model

Machine learning algorithms constitute the single most powerful generally applicable technology. They're also the coolest. By learning from data, they derive models that work—the models are capable of making predictions for new, unique cases. When training a model, the computer is essentially programming itself.

If you've excitedly jumped right to this chapter, then you're in good company—and yet you're exactly the person for whom I wrote the chapters that come before this one. You need to pay your business-side dues before you get to revel in this sexy rocket science. We all must fight our natural propensity to exalt the advanced tech in lieu of sufficiently obsessing over its launch—and in lieu of executing the preceding four project steps needed to make that launch possible. But if you've read all the pages before this one, you've earned the right to revel—enjoy!

This chapter delivers an accessible crash course for newcomers and business professionals to ramp up on ML algorithms. After all, these core methods have great ramifications: They will drive your large-scale operations. We'll keep it relevant by diving into the nifty principles of internal combustion, not how to change a spark plug.

The BizML Practice:

1. *Value*: Establish the deployment goal.
2. *Target*: Establish the prediction goal.
3. *Performance*: Establish the evaluation metrics.
4. *Fuel*: Prepare the data.
5. *Algorithm*: Train the model.
6. *Launch*: Deploy the model.

For the EduPay project, I was as happy as a pig in mud. I had 50 million rows of data, a viable business case for machine learning—ad selection—and a client willing to deploy.

But I also had my work cut out for me: I needed to generate models that would dynamically select between 291 ads in real time for one third of all college-bound high school seniors. That meant I needed to generate 291 models, one for each ad. With most ML projects, you only generate one model. Where to begin?

So much data, so little time.

As we've discussed, each model was to predict user response to its corresponding ad:

> **Prediction Goal for Targeting Ads (EduPay)**: Will the user respond to this ad if it is displayed?

I started by developing models for only a couple of ads. One had a lift of 3 at 10 percent. That means that the 10 percent of users predicted most likely to respond were three times more likely to do so than the average user. Those users might be the right ones to show this ad to rather than, say, the most universally popular ad.

The advertisement at hand recruited for the navy. It splashed the slogan "Accelerate your life" across a blue graphic and then asked, "<username>, are you ready to leave <city> and see the world? The Navy can show you how!" It presented two options, "Yes, please contact me!" and "No, thank you," followed by a "submit" button. The user had to choose one to get past this "interstitial" page and continue using the EduPay website to explore grants and scholarships for college. For each user who selected yes, the navy paid EduPay $12.50.

Let's dig into this model as an example.

Peering into a Model

When a newborn model emerges, it absorbs all your attention. Like counting a baby's fingers and toes, you examine it thoroughly, poking around to see how well it works and why—what makes it tick.

Algorithm 143

ML's discoveries are typically a mix of the arcane, inexplicable, and obvious. The navy ad's model included the following rule:

```
IF the user
    has opted in for marketing emails
        AND
    has not been shown this ad yet
        AND
    is in college
        AND
    has not specified a high school name
        AND
    has an SAT written over 480
        AND
    has an SAT verbal-to-math ratio between 0.5 and 1.5
        AND
    has an ACT score over 15
THEN the probability of responding to the ad is
    2.6%.
```

It may be hard to completely understand any logical rhyme or reason, but it worked. This rule—some call it a *pattern*—pinpointed a relatively responsive group of users. The overall response rate for this ad was 1.6 percent, so the users to whom this rule applied were 63 percent more likely to respond. In other words, it attained a lift of 1.63.

By discovering this kind of pattern, an organization takes "learning from experience" into hyper-drive. We've shown this ad to these kinds of customers in the past and the response was relatively good, so let's do more of the same. This transcends the traditional corporate move of simply doing more of what has been working, such as the Lands' End clothing retailer, which originally sold nautical supplies but noticed its clothing was selling well. ML goes further than that, providing more refined agility by eking out what has been working with precisely what kind of customer in precisely what situation.

The algorithm derived this rule from the data on its own. After pushing "go," I hadn't been actively involved in the process. It's like parenthood. In the immortal words of Forrest Gump, "My mama always said, life was like a box of chocolates. You never know what you're gonna get." Of course, Forrest's mama was talking about him.

Does the Model Make Sense?

Some parts seem obvious. Users who've opted in to receive marketing emails also tend to respond more to ads—not just this ad, but across all ads. No big surprise there. The same applies to whether the user has already been shown this ad. If not, they're more likely to respond since, if they're going to respond at all, they're mostly likely to do so the first time they see it. The lack of a specified high school probably corresponds to the fact that these users are already in college, so that part of the rule may mostly be redundant with the "is in college" and therefore inconsequential.

When you look into a model and see that it's mostly discovered things that seem obvious to you, that's no cause for disappointment. It means that your human hunches have been validated by the data. That validation is more crucial than you may realize. After all, what you don't see in the model are the many other potentially "obvious" but false presumptions that the algorithm has quite helpfully ruled out. Moreover, the model's value comes from not only the "obvious" discoveries but how it adeptly combines them together—along with some potentially surprising discoveries as well. By doing this well, modeling methods exhibit a certain logical and mathematical finesse.

Moreover, when a model seems to align with human intuition, that can help shore up confidence from business stakeholders. Depending on the culture and expectations of your organization, greenlighting a model's launch may depend in part on decision makers believing the model "makes sense."

Algorithm   145

But understanding a model is rarely straightforward, and whether it's important to do so makes for an unresolved religious debate across the ML industry. Take the rule shown above. It applies only for certain SAT and ACT scores and when the SAT verbal-to-math ratio is not terribly extreme—the user can be mentally lopsided, but not too lopsided. You can speculate on why such folks are more likely to respond to a navy recruitment ad. Perhaps military families tend to emphasize a well-rounded education, so avid recruits from these families are more likely to have more balanced scores. Or perhaps very lopsided minds are eccentric in some sense. The problem is, there's always more than one plausible explanation.

Having rules that predict is enough; understanding why they hold is both optional and unwieldy. Unless your data comes from a specially designed experiment, any interpretation meant to explain the reason behind a rule is no more than subjective speculation. To try to understand the "why" is to attempt to ascertain causation, which we cannot conclusively establish without collecting new data for that very purpose with a controlled experiment—thus the often-heard *correlation does not imply causation.*

Despite this, establishing causation is the whole point for many ML projects. This project's purpose was to pick the ad most likely to *cause* the user to respond. Certain ads tended to *cause* more responses for certain users, and the models served to capture such insights. To do this as well as possible, there's an advanced method called *uplift modeling* that holds advantages over this project's relatively simple approach of creating a separate model for each ad, but we didn't get that far in this project's execution and this book doesn't have space for it either. For more on that promising topic, see this chapter's notes at www.bizML.com.

But, broadly speaking, finding the causal explanation as to why certain inputs (e.g., "SAT verbal-to-math") link to increased customer response is outside the scope of business projects like this one. I wasn't working on a PhD in sociology in order to better understand human behavior. I was just trying to pick the best ad for each user. If the model predicts well and the numbers validate that beyond a reasonable doubt, do you care about the ultimate scientific explanation for the patterns it has discovered? Why should your company's decision makers require such explanations in order to trust a model that sheer numbers have already shown is reliable, especially when the attempted explanations are only conjecture?

In principle, decision makers shouldn't, but in practice, they often do. You certainly can't blame people for taking a look and speculating on the logic behind a model's machinations. Decision makers are often loath to greenlight a model without a look see. Otherwise, the model is a mysterious black box that many feel is hard to trust.

I can't resolve this religious debate and I'd be a fool to try. But even if you're dead set against deriving dubious explanations, there are other, definitive reasons to inspect each model. One is to check for bugs, which I turn to now. Another is to screen for ethical issues in how the model drives decisions, which I address in this book's conclusion.

Inspecting Models to Check for Bugs

With the EduPay project, I was grappling with a tough predicament: how to generate 291 models and then personally inspect each one, thoroughly.

For a typical ML project, you have to sanity-check each model. After all, there are many things that could go wrong, many potential gotchas and bugs in the data. I once had a model suggest that high-school dropouts were better hires. This prompted me to take another hard look at the data and how it had been obtained. There turned out to be a systematic problem in how humans had manually entered the data from the job-applicant resumes.

In another cautionary tale, researchers from the University of Washington generated a model that distinguished wolves from huskies within images. It showed great performance, but when they investigated how it was making decisions, they found it was actually operating based on whether there was snow in the background—the wolf images were more likely to have snow than the husky images. This

Algorithm 147

was a problem in the data, but it became apparent only when they inspected the model.

These things happen. In general, the data scientist must do some ad hoc poking around, performing a kind of model-integrity checkup. Before concerning yourself too much with any one model, the first question is whether the modeling method is fundamentally working, that is, whether the resulting model combines input variables in an effective way. Does the model mostly rely on only one input variable, failing to integrate other inputs as well? And if there is one overly dominant input, is its performance too good to be true—potentially revealing the kind of "time leak from the future" I discussed in the previous chapter?

At the same time, you look at the model's predictive performance, such as its lift. Does it pan out over held-aside *test data*, in comparison to how well it performs on the training data used to generate it? If not, this kind of underperformance is known as *overfitting*. It means the modeling process is memorizing particularities that are unique to the training data rather than truly *learning*—that is, finding insights that hold in general. If you're overfitting, you've got to troubleshoot how you've set up the modeling process or even just move on to another method.

On the EduPay data, a handful of models appeared to be passing muster. But unless I cloned myself or EduPay dramatically increased the project's budget, how could I scale this process to 291 models?

Before we get to my plan of attack for this project, let's take a step back and look at how modeling works in general and the wide range of modeling methods from which the data scientist may choose.

Learning from Data: The Ultimate Technology Challenge

After the first four project steps, we've completed the prep work and set the stage for the machine learning part of a machine learning project, the core technology itself: step 5, train the model, a.k.a. predictive modeling. The business case has been agreed on and the training data is ready. Now we get to feed it into the modeling software and push the "go" button.

When researchers set out to invent that software, they had their work cut out for them. They had to develop the step-by-step instructions for building a model from data. That is, they had to write a computer program that would generate a mechanism that itself would work with the dozens or hundreds of input values for an individual in order to calculate the predictive score for that individual. This model is expected to predict reasonably well for the example individuals within the training data used to guide its creation. More importantly, it must also predict well for a held-aside *test dataset* of examples. This separate set of cases serves to estimate the model's performance in general on unseen cases never before encountered. That evaluation step tells you whether the modeling has succeeded.

What a challenge! Imagine the plight of your poor computer, a feeble, knowledgeless mechanism. You pour reams of data into it, but it doesn't "understand" the data. The variables have no meaning to it. It has no general knowledge about user profiles and behavior as you do—nor does it grasp what it's trying to predict, the real-world meaning of the output variable.

The model has to take into consideration all the dozens or hundreds of factors known about a new, unique situation or individual. Given the input variables that define that situation, how should it weigh or combine all these factors to calculate the most precise probability of a positive outcome? More to the point, how could the computer automatically learn to do that—that is, how could it automatically generate that model?

You're about to find out. In the face of this ultimate challenge, predictive modeling methods such as those described in this chapter achieve scientific greatness: Their models work. The generalizations drawn from past examples still hold when applied to new, never-before-seen situations. This capability makes ML the world's most powerful widely applicable technology.

Modeling methods are tried and true, born of research labs and proven in commercial deployment to be adept and robust. When we

Algorithm   149

use them, we're standing on the shoulders of giants, the researchers who developed them. And in beholding them, we're enjoying the privilege of seeing what methods have turned out to work best, skipping past all the trials and tribulations that those inventors suffered. As with genius, some say that research is 1 percent inspiration and 99 percent perspiration. But for a commercial user of established ML methods, it's no sweat—or at least a lot less sweat.

It's easy to get excited by how profound an endeavor this is. Let's connect that excitement to the concrete mechanics by looking at exactly how it works.

Decision Trees: Models Made of Rules

One of the most popular modeling methods is *decision trees*. A decision tree is made up of if-then rules like the one we looked at above—which I did indeed extract from a decision tree. Here's an example:

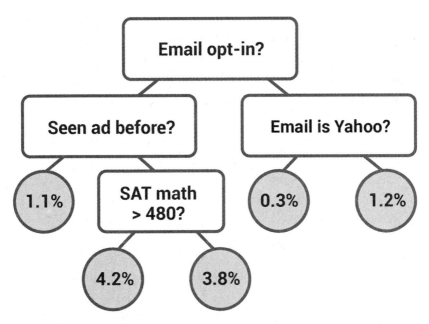

A decision tree to predict ad response. Start at the top. If the answer is yes, go left; otherwise, go right.

The modeling process automatically creates a tree like this from the training data—usually ending up with one much bigger than the example shown above. Then, to use it in deployment to predictively score an individual (e.g., an EduPay user), you simply start at the top (the *root* of what is an upside-down tree) and by answering yes/no questions, you make your way down to an end point (a *leaf*) to derive the score for that individual. For example, if the answers are yes, no, yes, then by going left, right, and then left, you wind up at a leaf with 4.2 percent—the model is saying there is a 4.2 percent probability that the individual will respond if shown the ad.

You can think of a decision tree as a bunch of nested if-then-else statements (if you've done some programming), as a flowchart with no loops, or as a bunch of rules—each path from the root down to a leaf makes a rule. For example, for the path we just covered, the rule is: *If the user has opted into email and has not seen the ad before and has an SAT math greater than 480, then the score is 4.2 percent.*

The way modeling builds a decision tree is to "grow" it from the top down. It starts with the most predictive single input variable at the top, thereby dividing all individuals into two groups: in the case of the example tree shown, those who've opted in for emails and those who have not. Then, it further subdivides these groups as it builds the tree in a downward direction. This repeats as the training data is divided into fairly small groups, although not too small, since you can't generalize well from only a handful of examples. By following this process, the tree's size and shape—and the choice of input variables for its yes/no questions—are all determined automatically.

More Modeling Methods: Linear and Logistic Regression

Other modeling methods look completely different and work completely differently. *Linear regression* creates a *linear model* that simply combines input variables with a weighted sum, such as:

(0.0008 × SAT-written) + (0.4 × email-opt-in) + (0.16 × in-college)

Algorithm   151

This example linear model—which involves only three input variables—assumes that the *email-opt-in* and *in-college* variables have the value 1 when true and 0 when false.

In this case, the job of the modeling method is to adjust those three weights, tweaking them until the model does as well as possible on the training data. It's unlikely to do especially well; this old-school, standard statistical method has become overshadowed by modern modeling methods.

However, linear models serve as the basis for other methods. One such method, which is very popular, is *logistic regression*, which is simply a linear model followed by a *nonlinear transformation* known as an *S-curve* or *sigmoid function*. This extra step "stretches" predictive scores in the mid-range to be closer to 100 percent or 0 percent probabilities, in effect trying to get the model to commit to more definitive "yes" or "no" predictions. This turns out to do a better job for many yes/no prediction goals (a.k.a. *binary classification*) than a plain linear model.

Now, if you're not a data scientist, you might wish you could skip over these technical details for the most part. Allow me to grant your wish.

Everything You Need to Know about Modeling Methods

Before we briefly survey several more modeling methods, let's get one thing clear: As vastly different as they may seem to be from one another, all models accomplish the same simple task: They take the inputs and output a predictive score. That's literally the definition:

> **Predictive model**: A mechanism that predicts a behavior or outcome for an individual, such as click, buy, lie, or die. It takes characteristics of the individual as input (input variables) and provides a predictive score as output, usually in the form of a probability. The higher the score, the more likely it is that the individual will exhibit the predicted behavior.

Since the model is generated by ML, we say it's the thing that's "learned" or "trained." Because of this, ML is also known as *predictive modeling*.

Earlier in this book, we depicted a model as a black box—more precisely, a golden egg:

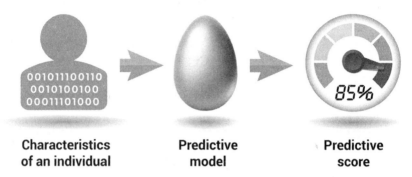

| Characteristics of an individual | Predictive model | Predictive score |

Scoring: A model generates a prediction for an individual.

Inside that egg, if it's a decision tree, then it's applying rules to derive the score. If it's a linear model, it's applying arithmetic. But whatever it's doing on the inside, the model is always used in the same basic way: The organization retrieves scores from the model and guides operational decisions accordingly, irrespective of how the scores were calculated.

From that utilitarian perspective, you can conceive of a model as a black box that outputs predictive scores. To make use of those scores, the model's inner workings make little difference. In fact, the other five bizML steps outside this one operate in mostly the same way, regardless of which kind of model your data scientist develops. Those steps can be executed with little concern about the choice of modeling method, so long as the model does its job without consuming too much time or computational resources.

Even so, the data scientist who owns this model-training step doesn't have the luxury of skipping past the inner workings. Why must they suffer the headache of a potentially endless range of different methods, each with its own arcane technical details? Why can't there just be one best method?

Algorithm   153

Why There Are Competing Modeling Methods

There is no one best *predictive modeling method* (a.k.a. *ML algorithm*). There's no "Holy Grail," no universal champion. No matter how good a modeling method is, there are always some datasets that would be better handled by another method. The method that performs the best depends on the problem at hand—ultimately, on the nature of the data.

As the field of ML develops, it's not converging toward a single best modeling method. This fact is encapsulated by the *No Free Lunch* theorem, a famous principle among those on the more academic and theoretical side. It's a cute name that makes the point: If there were one method that always wins, the data scientist wouldn't have to work as hard. It'd be like receiving a free lunch.

Human nature will sometimes try to defy the undeniability of No Free Lunch. ML practitioners often become quite infatuated with one modeling method or another. But the fact is, you never know what you might be missing by not trying an alternative method until you try it out on your data.

The diversity of solutions is a good thing. The industry thrives by leveraging a wide range of competing methods. Even if it were theoretically possible to discover one winner-takes-all algorithm, that would be the wrong wish to ask from a genie. Each method has its own pros and cons—not only with regard to predictive performance, but also regarding all kinds of pragmatic factors such as speed, understandability by humans (a.k.a. *model transparency*), complexity, and the level of human expertise required to use it.

The diversity of methods stems from their diversity of conception. Today's wide range of methods were invented at different times, in different countries, by different people. Each algorithm was conceived as and designed around a researcher's intuitive notion, something that seemed like a good idea to that person.

Human ideas tend to be relatively simple, but that helps ML work—because *simplification facilitates learning*. Modeling methods only work because they oversimplify. Learning from examples is only possible by way of an "inductive leap," a simplifying assumption about the world. For example, decision trees impose a very simple structure: rigid if-then rules based only on whatever limited set of input variables has been made available. Even if it were theoretically possible to predict perfectly, it certainly wouldn't be accomplished with that kind of simple mechanism. However the world works, it could never be expressed in such simple terms. But although far from perfect, the patterns expressed by those if-then rules do indeed pan out in general, providing a lift over pure guesswork. Modeling methods only work—they only succeed in drawing generalizations from examples—because of the limitations in how patterns can be expressed that this kind of structure imposes. Without any such structure, an ML algorithm would overfit the data, memorizing its peculiarities rather than gleaning insights that hold in general.

Having such a diverse range of methods empowers the data scientist. She has a versatile, dynamic toolkit of options at her disposal. Depending on the project's particulars and requirements, she can try out various options, following her gut and the pragmatic rules of thumb that she has picked up throughout her career. There's an ad hoc aspect to this process, but, with experience, experts gain a sense for how to proceed.

This range of available methods is especially valuable when you combine them together, having them cooperate rather than compete. An *ensemble* of diverse models itself is in fact one of the most important kinds of modeling methods. Let's see where it fits in among a short list of popular methods.

A Summary of Modeling Methods

There's a certain commonality among today's diverse hodgepodge of modeling methods. As we've discussed, they all accomplish the same thing: Generate a model that takes input variables and produces a predictive score. Moreover, they all do so in a similar way: Start with a

Algorithm   155

crummy model—one that is very small or totally random—and then iteratively tweak it, repeatedly making small modifications to it so that its performance across the training examples improves.

Since each incremental improvement corrects the model so that it gets some cases right that it had been getting wrong, you can think of the process as a regimented, automated way to improve a scientific hypothesis or theory (in fact, in past decades some ML researchers used the word *hypothesis* instead of *model*). As Clayton Christensen put it in his book *The Innovator's Dilemma*, "The key to improving any theory is to surface anomalies—events or phenomena that the theory *cannot* explain. It is only by seeking to account for outliers—exceptions to the theory—that researchers can improve the theory." In a way, modeling automates the process of refining a hypothesis.

Modeling methods are supremely general-purpose, applying across industries and organizational functions. Each kind of model can serve marketing, financial risk management, fraud detection, or clinical healthcare—what makes the difference is the data you give it. With all the concrete ways in which we apply them across domains, the methods themselves are designed in the abstract. The input and output variables you pull together as training data determine whether the model will predict sales or successful surgeries.

Here's a summary of some of the most popular modeling methods (these all depend on labeled training data, which makes them *supervised* ML methods):

	What's learned from data during model training	Once trained, how the model generates a score	Pros and cons
Decision trees	The decision tree's architecture: its size, shape, and choice of inputs	Start at the top (the root) and flow down to an end point (leaf).	Easy to interpret (transparent) and surprisingly effective for its simplicity, although usually outmatched by more advanced methods.

	What's learned from data during model training	Once trained, how the model generates a score	Pros and cons
Logistic regression	A weight for each input	Apply the formula to the inputs: Add up a weighted sum of the inputs and then apply a nonlinear adjustment.	Easy to interpret, but usually outmatched by more advanced methods.
Naive Bayes	A factor for each input for positive cases and the same for negative cases	Apply the formula to the inputs: Roughly speaking, multiply the inputs' factors for positive, then for negative, then normalize.	Easy to program and robust against overfitting but limited in predictive performance.
Ensemble models	A set of simple models—sometimes all decision trees (e.g., *random forests* and *TreeNet*) and sometimes varied (e.g., *boosting* and *bagging*)	Score with each simple model and then combine the scores, e.g., by averaging them or taking a vote.	An elegant way to improve over simple models, but the resulting amalgam of models is difficult to interpret (opaque).
Deep learning	The many weights within a large, complex mathematical formula (a *deep neural network*)	Apply the formula to the inputs (complex).	A breakthrough advanced method, which can handle a great number of inputs—e.g., each pixel of a high-resolution image—without the need for preprocessing, but difficult to interpret (opaque), computationally expensive, and often requires highly technical human expertise to use successfully.

When experts mention one of these, such as decision trees or ensemble models, they are referring to two things at once: the structure of the model and the process to train it. These two aspects are always paired. The model's structure can be things like a tree, a simple formula, or a

Algorithm   157

complex network (which is just a way to visualize a complex formula). And then, for each type of structure, the process to train it—to perform modeling—is specialized for that structure.

In some cases, the modeling *process* has its own name, separate from the type of model. For neural networks, the method is *backpropagation*, which tries the model on a training case and, to the degree the output is wrong, propagates negative feedback backward through the network, adjusting weights accordingly. By doing this repeatedly, the model's performance improves. Originally developed in the 1980s, backpropagation still trains today's more complex and capable neural networks, *deep neural networks*. This method is called *deep learning* (named in this case after the modeling process rather than the model's structure).

Why Are Modeling Methods Also Called *Algorithms*?

What do you call a formula that can predict Al Gore's dance moves? An Al Gore Rhythm.

—Anonymous

Modeling methods are also called modeling *algorithms* or ML *algorithms*—because *algorithm* simply means a process to get something done.

Algorithm: A well-defined, finite process that solves a problem.

In the context of ML, *algorithm* refers to a modeling method, such as decision trees or logistic regression. But more generally, algorithms are fundamental to the whole of computer science. In practice, the word just means any procedure that is defined specifically enough that you could program a computer to do it. The word may sound technical, but it's a simple and intuitive notion.

The way media has come to use the word *algorithm* tells us a lot about ML's elevated status. Even though the concept applies for any and all of the many things we do with computers, the press uses *algorithm* as a synonym for ML. It's like how hair stylists use the word *product* to mean hair product. Or how the word *crypto* usually refers to cryptocurrency even though *cryptography* applies to securing any and all kinds of transactions and communications. When a field gets famous, it dominates more than its fair share of vocabulary.

Choosing a Modeling Method

Better prediction doesn't come for free. Broadly speaking, the better the method, the more complex—both to use and to interpret how the resulting model works. You can see that progression as you go down the list of methods within the table presented earlier. After the first three simpler types of models—decision trees, logistic regression, and Naive Bayes—you come to a more complex, adept one: ensemble models. To improve on simpler models, an ensemble literally "ensembles them together"—so by definition, ensemble models are more complex. By the end of the list, deep learning, you've increased capacity greatly for certain problems, but you've also piled on complexity.

For many projects, the *interpretability* you get with simpler models is a godsend. As we've discussed, being able to understand what makes a model tick is sometimes critical for debugging and for convincing certain decision makers to fully buy in. And auditing a model for ethical considerations—to inspect how it renders consequential decisions that affect people—also hinges on understanding its innards. This desirable model characteristic is also known as *explainability* or *transparency*. It's at least somewhat lost when you move to the impenetrable, unwieldy soup of math you get with more complex models. The long if-then rules of a decision tree may seem arcane and difficult to understand, but on the scale of things, they're relatively friendly to human eyes.

In the name of interpretability, UPS kept it simple for its package-delivery prediction. To form predictions for tomorrow, the system looks at how often each address has received a shipment on similar days, such as the same day of the week, or, in some cases, a more specific day, such as the day after Thanksgiving. Then, the probability for each address is adjusted by an overall forecast. For example, if on the whole more deliveries are expected tomorrow than average, then each individual destination's probability is adjusted slightly upward so that the aggregate expected count is consistent with the forecast. The effect of this scheme is similar to—and has the same simplicity of—a decision

Algorithm   159

tree. With a model so straightforward and intuitive under the hood, project staff members had an easier time selling its deployment across the company.

Another reason to stick with simpler models is that even if you pay the extra price of complexity, it won't necessarily improve prediction. More complex methods aren't guaranteed to do better. Sometimes it isn't worth it to go "full rocket science." Again, the best method for any given data just depends. There is no one method sure to always do best. Every ML project has an experimental aspect to it: You can only know how well a method will work by trying it.

When it comes to choosing a method, human judgment is key, since rigorously comparing every competing method is impractical. It's usually possible, for example, to give all five of the methods listed in the table above a preliminary shot, or even to try out three or four times as many, if you invest a lot of time. The challenge is that, for each method, there are many settings that affect its performance—too many to exhaustively test. To address this, techniques from the field *AutoML* serve to systematically explore many methods and settings. AutoML is an increasingly popular, albeit inexact, approach. It helps scale your search for the best modeling setup, but it does not replace the expertise of a data scientist.

Ultimately, a rigorous head-to-head comparison between even just two modeling methods is a challenge to accomplish, since each one has so many knobs and dials with which you could experimentally fiddle ad infinitum. Champions of any one method—data scientists who are fans and have become highly experienced with that method—tend to get the best performance out of it. Given this human factor in model performance, the conclusiveness of any head-to-head trial pitting methods against one another can always be questioned.

In the end, the data scientist must be artful. Their experience and intuition guides the choice of model. Practically speaking, for many ML projects, data scientists end up trying out only a few methods. Besides, there's another, better opportunity for improving predictive performance.

It's the Data, Stupid!

> We don't have better algorithms than anyone else. We just have more data.
>
> —Peter Norvig, director of research, Google

> The summation of my decade of data work is that the data matters more than the model. Every time.
>
> —Caitlin Hudon, principal data scientist, OnlineMedEd

Fine-tuning an algorithm and trying new ones for comparison will only get you so far. As you put in more and more effort, the returns tend to diminish. You might make small, incremental improvements by stepping up to a more complex model like a big, hairy ensemble or even deep learning—but it's often not worth the complexity. Sometimes, you tweak the heck out of your modeling efforts for only a small gain. On other occasions, for certain projects where every little improvement makes a big difference, it could be worthwhile to go the extra mile. Even so, that extra effort may provide only a small gain and may pay off only for a few brief months before changes in the business require a full project restart.

Either way, the mandate is clear: Improve your data. That's where your efforts will usually pay off much more handsomely. This means both more data and better data. Google research director and famed *Artificial Intelligence* textbook coauthor Peter Norvig espouses getting more data, but that's not all. "More data beats clever algorithms, but better data beats more data," he says.

Bettering data quality rather than its quantity is harder to formally define, yet it's just as intuitive. Just think about what might make data more predictive. In the previous chapter, we discussed how to design more revealing, informative inputs through a *feature engineering* process. For example, FICO developed a summary of cardholder propensities and changes in spending behavior. As another example, a telecom expanded its churn model inputs by merging in summaries of how a

Algorithm   **161**

user tends to use the website. Cellphone subscribers who'd checked out their remaining contractual obligation were more likely to cancel. Beyond adding new, informative inputs, one can always exert more effort in data quality assurance, seeking out where there might be errors.

But this takes discipline. Only the most regimented ML leaders focus on this "predictive fuel" more than on the sexy "prediction engine." As Google ML software engineer Josh Cogan puts it, "I find most people really tend to focus just on optimizing the ML algorithm. They want to make sure they have the newest, coolest thing right out of the papers. . . . I've never found anyone who overestimated how hard it was going to be to get that data collection right in the first place."

Also at Google, a group of six researchers put it on the line with a plea for sanity: "Data quality carries an elevated significance in high-stakes AI due to its heightened downstream impact, impacting predictions like cancer detection, wildlife poaching, and loan allocations," they wrote in a research paper. "Paradoxically, data is the most under-valued and de-glamorized aspect of AI." They titled their paper so that nobody could miss the point: "Everyone Wants to Do the Model Work, Not the Data Work."

To emphasize data's importance, data science has borrowed a clarion call from political science. "It's the economy, stupid!" proclaimed political strategist James Carville to campaign staffers working in 1992 to elect Bill Clinton. By adopting that as a campaign slogan, they won the presidency. In more recent years, ML has revised that legendary catchphrase to be its own mantra: "It's the data, stupid!"

And yet data isn't the only game in town. More complex models do have their place for certain projects.

How Deep Is Your Learning?

The latest, greatest wave in modeling methods, *deep learning*, produces models that involve more complexity than ever before. And it often pays off. Deep learning is an advanced form of *neural networks*, which represent complex mathematical formulas as networks organized into layers.

Deep learning has improved the state of the art in modeling so that it can truly take advantage of a large, complex model. It's always been the case that you could set up a neural network with many, many inputs—it's just that it never used to work well. For example, if you want the model to detect whether a 1280 × 720-pixel resolution image includes a traffic light, you could directly feed it all 921,600 pixels. That's a hugely greater number of inputs than the typical few hundred provided when, say, predicting customer churn. The problem is, for the model to handle each detailed image in a useful way, it must be possible to train it for adept, complex processing. Not only must the model be complex, but the training must be capable of leveraging that complexity. The original algorithms a few decades back always flopped when this was attempted.

Because of this, neural networks, which originated in the 1980s, started small. Back in 1997, the first time I taught the graduate ML course at Columbia University, neural networks were shallow rather than deep, and yet, for many domains, they were nonetheless the leading option. They were already steering self-driving cars, in limited contexts—but the input was only a very low-resolution view of the road ahead, an image of 30 × 32 pixels. I even had my students apply neural nets for face recognition as a homework assignment, using the same low resolution.

Increasing the layers of a neural network—so that it is literally deeper—increases the complexity of the mathematical formula that it embodies. But, for decades, the modeling process (backpropagation) was incapable of scaling up to take advantage of that complexity. Since they couldn't make use of the additional layers, neural networks couldn't effectively handle a large number of inputs.

As data exploded, so did advanced modeling. After the turn of the century, amazingly, modeling methods improved to meet the challenge—and realize the potential—of properly training deep neural networks. This was possible due to bigger data and faster compute, as well as some new improvements to the modeling algorithm itself.

Deep learning blossomed, both in buzz and in proven value. It is the one and only technology that achieves and defines the state of the art for speech recognition and for various image-processing applications,

Algorithm   163

including diagnosis on medical images, autonomous vehicles recognizing and classifying objects around them, and everyday operations like unlocking your phone with facial recognition.

Google, for example, has made significant improvements to most of its main products with deep learning, including Android, Apps, Maps, Speech, Search, and YouTube. Gmail now intercepts 99.9 percent of spam. And your unlabeled Google Photos are searchable by ad hoc terms such as "hug." Google Translate—which anyone can use online—swapped out the original underlying solution for a much-improved one driven by deep learning. Go try it out—translate a letter to your friend who has a different first language than you. I use it a lot myself.

The exploding field *generative AI* is built on deep learning. This includes *large language models*, which write prose that is impressively coherent for a computer; *digital image generators*, which take a written prompt and create an image for you; and *deep fakes*, depictions of people who do not exist or of existing people doing things they've never done. Generative AI also produces synthetic music, speech, and video.

To keep up, we launched the Deep Learning World conference in 2018 as part of the Machine Learning Week conference series that I founded in 2009. They take place annually on two continents.

As much as I can't help but geek out like mad over deep learning, here's the thing: This all-powerful Hulk isn't the right superhero to enlist every time. When you need someone to crawl into a mousehole, Ant-Man would be a better choice.

For Many Business Problems, Deep Learning Is Overkill

Deep learning tends to solve different types of problems, in comparison to so-called classical ML algorithms. For one thing, it's more often used for detection rather than the prediction of the future. Since it takes so many inputs, a deep learning model can directly handle an entire unprocessed file, such as a photograph, medical image, or an audio file for speech recognition. This lends itself to detecting whether each contains a certain element or belongs to a certain category.

This tends to pigeonhole deep learning into certain industries, those where you need to do detection on raw image and sound files, such as for autonomous vehicles, medical image processing, and devices that need to recognize speech. That's a somewhat distinct arena from the kinds of customer-prediction applications that this book largely focuses on, such as for targeting ads and marketing, managing financial risk, averting fraud, and predicting which locations will receive a package delivery.

Deep learning applications also tend to allow for high accuracy—in the casual sense of the word: *the ability to classify correctly a great majority of the time for both positive and negative cases.* Just as humans can usually tell which photographs do or do not include a traffic light and which sound bites do or do not include the word "hello," so too can deep learning.

In contrast, customer prediction, typically the domain of classical ML methods, means predicting what people will do. For those applications, you can only hope to gain some meaningful lift in comparison to guessing, unless you have a magic crystal ball. Not even deep learning can confidently predict human behavior in general. No matter how sophisticated and advanced the model and the modeling algorithm, it doesn't change the fact that we're trying to predict the unpredictable: people.

Given that there's an intrinsic upper limit in predictive performance, deep learning is usually overkill for many of the more classical, standard business applications of ML. For those problems, deep learning's awesome capabilities are often wasted, and working with a classical ML method would do virtually as well, requiring less complexity, time, computational resources, and advanced expertise. Moreover, simpler methods maintain transparency, so it's possible to more easily understand the model itself, to see what's been learned and on what basis predictions and decisions are being made by the model.

On the other hand, there are exceptions where deep learning has provided value for traditional business applications. For example, FICO's Falcon fraud detection system is based in part on deep neural

Algorithm   165

networks. The research and advisory firm Celent has estimated that, if more widely deployed, deep learning could reduce fraud losses by $161 billion worldwide, across all kinds of financial fraud.

"Deep learning largely dominates today's media buzz about machine learning," points out industry leader Dean Abbott. "It sucks the oxygen out of the room."

Still, deep learning constitutes only a subset of the field as a whole. Like all modeling methods, it belongs within one corner of the taxonomy of techniques.

Despite its unique nature, deep learning doesn't impose much change on the project's business-side execution. Successful deployment requires the same six-step bizML practice. You're still altering operations with predictive probabilities by integrating them into existing systems. You still need to ramp up decision makers and line of business managers. You still need to surmount the universal challenges of data preparation. And the organization still needs to understand and agree on quantitative measures of predictive performance and business performance. It's a different animal, but the animal trainer's skillset translates nicely.

Machine Learning Software: How to Choose a Tool

So many powerful ML algorithms, so little time. To make use of these algorithms, you need software that implements them. The good news is there are many competing solutions out there. For more than two decades, I've noticed new ones come across my radar every two or three months. But this exploding plethora of options is enough to give you analysis paralysis, with an incapacitating fear of buyer's remorse. How to proceed? Here are some tips and pointers.

Don't program ML software from scratch. Buy rather than build, download rather than develop. ML projects almost always leverage existing analytics software. All the standard predictive modeling methods are each implemented within many software tools that you have at your disposal. Even the latest, cutting-edge techniques straight out of the research lab have often been released as open source code by the researchers themselves. For industry ML projects, circumstances only very rarely warrant programming the algorithm from scratch.

Software selection doesn't guide the project—project requirements and team skills come first. Do not hold any illusions that an analytics software tool can be a "plug-and-play" solution to the business problem you're solving. The goal of an ML project is a new way of business, an improvement to operations, for which ML software plays a central but limited part. Some ML vendors may offer an aggressive sales pitch that you will need to resist. Instead, allow your data scientists to determine software requirements as the project matures and evolves. If your data scientists have a strong preference or are already well versed with one tool in particular, their work may be most effective using that tool.

Postpone the decision. If no preordained tool is in place, don't ordain one until absolutely necessary. After you've greenlit a project and proceeded through the first four steps—through data preparation—then you will have pertinent data on which to evaluate modeling software. Many vendors will provide a free evaluation license, so you can give it a go on your data, possibly comparing multiple tools. Moreover, other determining factors will also have become clear during later phases of the project, including your budget to buy software, how its models must integrate with your existing systems and data pipelines, and whether deployment is to be cloud-based or on-premises.

Use what you've got. If your organization has already adopted a solution, take a good hard look at that product as your first consideration. Often, a team will end up adopting a combination of paid and free, open-source tools, many of which play well together.

Generating Numerous Models for Ad Targeting

Let's return to the challenge I faced for EduPay. I had to generate 291 deployment-ready models, somehow streamlining and scaling the quality-assurance process by eliminating the need to spend a lot of time poking around and inspecting each model.

A couple of months into the project, I had an "aha!" moment. I ran a few experiments and then emailed my client to propose a solution:

Algorithm   167

Melissa,

I completed a few more rounds of modeling with a different method: *Naive Bayes*. The results are good—here are the takeaways:

- Naive Bayes gives us complete coverage. It would enable us to churn out a massive number of models—one per ad—in "one fell swoop" with the push of a button and little or no manual "fiddling" for each individual model.
- Naive Bayes did almost as well as decision trees with little "fiddling" on my part.
- Naive Bayes is free (I programmed it from scratch).

Here's a brief description of how Naive Bayes works: It calculates a predictive "degree of evidence" for each input. For example, say a user is in tenth grade and is in ROTC and those both provide a degree of positive evidence that the user is more likely to respond to the ad. Those degrees of evidence are aggregated by simply multiplying them together. There's some probability theory, but the main part is multiplying together the degrees of predictive evidence that correspond with the input variables.

Let's discuss further!

—Eric

Because of its simplicity, Naive Bayes works reliably. It's robust against overfitting because it doesn't try to draw generalizations from small groups of individuals. In contrast, a decision tree drills down to "sub-sub-segments," such as *all tenth-grade males who are not US citizens and have Hotmail as their email provider.* When you look at such a specific group, it may include only a handful of cases. But Naive Bayes considers only one input variable at a time, never drilling down to such specific, smaller groups.

With this reliability, I wouldn't have to inspect each model to check for problems. As expected, each model's performance held up on unseen test data. And I was able to do enough checking for data issues or other bugs just by inspecting several models.

With Melissa's approval, the project proceeded—and yet I was breaking the first rule from the sidebar above about ML software: *Don't program ML software from scratch.* This unusual situation was the exception that proves the rule. At the time, there weren't a lot of off-the-shelf options specialized for personalized ad targeting with a model. By programming it myself, I could implement it right inside EduPay's existing database, without learning about, choosing, adapting, and integrating someone else's product.

In the time since this project, the operationalization of ML has come a long way and you're much more likely to find existing software that's a match for your project. My coding from scratch was a singular exception, the only time I've done so, other than during my previous life in academic research. The same track record holds for all my data science colleagues, to the best of my knowledge.

With the model training complete, it was launch time. I now had to work with a company new to predictive models, guiding them to integrate the models into the heart of their principal operations. What would it take to navigate this final mile of the project? The next chapter answers this question and completes the EduPay story—but first, it recounts a prequel episode from the UPS story about a time when deployment went awry.

6 Launch

Deploy the Model

To deploy a model is to propel it from the lab to the field where the enterprise will drive operational decisions with its probabilistic scores. To be specific, each time the model scores an individual, that score directly informs the action taken for that individual, such as whether to contact, approve, or audit. This culminating project step is where machine learning begins to deliver its value (so begins *model upkeep*, which is covered in the conclusion).

Deployment requires full-stack organizational buy-in, cooperation from staff at every level. While executives are the ones who approve it, operational staff must also agree, since that's where deployment introduces change. How do you overcome the resistance to such change? How do you mitigate the risk of deployment mishaps and assuage even the most risk-averse stakeholder? How do you translate predictive probabilities into operational actions? How do you engineer the right data into the model on the fly and get the model to work in only milliseconds for high-speed processes? In the end, even if performance improves, how do you prove that the credit should go to your ML project rather than to other changes within or outside the organization? As UPS and EduPay conclude their stories, we'll have our answers.

The BizML Practice:

1. *Value*: Establish the deployment goal.
2. *Target*: Establish the prediction goal.
3. *Performance*: Establish the evaluation metrics.
4. *Fuel*: Prepare the data.
5. *Algorithm*: Train the model.
6. *Launch*: Deploy the model.

After step 6: Maintain the model (covered in the conclusion).

Jack Levis was in deep water at UPS. So far, trial runs of Package Flow Technology—the system that deployed his delivery-prediction model—had only delivered disappointment. "Things were really ugly internally," Jack reflects. "It was a nightmare."

The rocky road to deployment will test any technology pioneer's mettle.

But this wasn't only an internal affair. The media had caught wind of it and blown the lid off. "New Package Flow Technology Not Delivering at UPS," screamed a *Computerworld* headline. The feature story continued, "Its highly touted Package Flow Technology isn't flowing as smoothly as expected, with problems at about a third of the 300 or so centers where it has been implemented."

Within an office shut off from the kerfuffle, the COO of UPS chastised Jack privately. It was a heated discussion and the fallout would reverberate for a long while. Even two years later when they were discussing the next project iteration, the COO looked Jack dead in the eye with the stare that only a Fortune 500 executive could muster. "I don't want another Package Flow—don't you dare do that."

Remember that stiff upper lip Jack had perfected early in this book? It comes in handy when weathering the hailstorms that arise when managing organizational change.

But Jack had good reason to defend his innovation: The problems so far weren't in the technology—they were in the humans. To deploy delivery prediction at UPS was to ask people to change their habitual routines and embrace a new paradigm. It's a story as old as machine learning: The deployment plan was easier said than done.

Shift Happens: When a Legacy Process Goes Digital

Out with the old, in with the new. Package Flow Technology (PFT) was designed to boost efficiency by replacing each shipping center's legacy process with one that's more automated and centralized:

Legacy process: Each day, humans assign the delivery regions (*sequences*) that each truck must cover. Many of these decisions come while loading the trucks, during which time the staff adjust the assignments in an ad hoc manner as they deem necessary. This sometimes means reassigning packages that have already been loaded, shifting them from one truck to another.

Updated process: The PFT system centralizes and semi-automates the assignment of sequences to trucks, based largely on predicted deliveries. Just before truck-loading begins, a planning manager completes final adjustments through a central PFT console in the hope that little to no further revisions will take place on the fly during the loading process.

If adopted fully, this process change would radically improve the efficiency of operations: It would decrease the mileage—and the time clocked by drivers—accumulated across the entire fleet of trucks. It could accomplish this because of two fundamental advantages over the legacy process. First, it dynamically incorporated the prediction of as-yet unknown deliveries in order to plan and begin loading the trucks early for on-time departures. Second, it centralized decision making so that it applied across all the shipping center's trucks at once. This would beat the legacy process's distributed decisions made by individual truck loaders on the fly while loading.

With the PFT system in place, these two advantages held even when managers made manual adjustments to the plan. When they did so, it was at a central console, with a bird's-eye view across all the trucks going out that day. The console incorporated the day's predicted deliveries along with known deliveries. As a manager revised the plan on-screen, it displayed the forecasted effect based on both known and predicted deliveries.

But the system had some "bugs": the humans—in particular, those who were carrying out its instructions. If the staff loading the trucks overrode the centralized decisions too often, the benefits of PFT and delivery prediction would vanish. Changing a package's truck assignment meant not only enacting a potentially suboptimal decision without the bird's-eye perspective provided by the central console; it could also mean inefficiently moving packages that were already loaded. This risked delaying trucks so they wouldn't depart on time. Moreover, when staff vetoed the system and loaded a package onto another truck, they typically wouldn't update the system. This meant the physical world and the digital world didn't align—and that spelled trouble. Following the data on their handheld device, a driver would go to deliver a package that was in actuality on another truck, and the driver of that other truck wouldn't even know they had the package. To address this misalignment, Jack formulated a new mantra for his staff: "The data is as important as the delivery."

Clearly, Jack's team had more work to do getting these staff members to change their ways. With the legacy process, staff had applied their hard-earned knowledge and experience. If a seasoned truck loader saw a package with a delivery address that they recognized, they'd reflexively say, "Oh, that's got to go on the truck with this other package." To realize the potential gains in efficiency, Jack and his team would have to convince and reorient staff to more strictly follow a preordained plan while loading the trucks.

To Manage Change, Change Management

Two-thirds of my effort was deployment, versus models and build with IT.

—Jack Levis

Staff at every level resist change, from the loading dock to the top-floor offices. Recall another precarious moment from earlier in this book, when Jack, struggling to gain authorization from Chuck, the executive, literally took him for a ride to demonstrate the system's truck navigation. Just as Chuck initially had a tough time swallowing decisions that seemed to defy his human intuition, so did those who loaded the trucks.

But staff at every level must cooperate. Deployment stalls when an organization can't or won't. To gain buy-in from both the bottom and the top is to ensure both the "can" and the "will" of deployment—both the capacity and the authorization. *Executives* approve change, and yet those *executing* must also comply. We must achieve full-stack organizational buy-in.

Different decisions, same story. This chapter's story is a prequel to the episode with Chuck. Here, Jack works to get prescribed truck *loading* implemented properly. Only later did he get Chuck to approve prescribed truck *navigation*.

The efficacy of technology so often comes down to human adoption. "The most difficult part of my job is not actually working with mathematicians to come up with a beautiful model to solve a problem," lamented a keynote speaker years later, as he strode across the stage at Machine Learning Week 2022. As the conference chair, I'd enlisted another leader from UPS, network planning and optimization director Yentai Wan. He continued, "The most difficult part of my job is actually deployment. It's the so-called change management. How do I convince those end-users to switch from the legacy system and leverage the modernized technology we build out?"

Such change-management challenges aren't new in general, but when it comes to ML projects, the need to shrewdly manage operational change is often overlooked. The advanced modeling algorithm itself absorbs much of the project's attention and seems to promise the moon. ML delivers a rocket, but those in charge still must oversee its launch.

Training Daze

Large-scale change requires advancing an inspirational vision, building relationship capital, and maintaining organizational alignment . . . leadership simultaneously embraces unifying and disruptive ideals.

—Christopher Hornick, *The Last Book of Leadership*

> In order for transformation to be successful, leaders must approach
> it in ways designed to . . . drive emotional commitment from
> employees.
>
> —Andrew White et al., "Organizational Transformation Is an
> Emotional Journey," *Harvard Business Review*

With PFT's effectiveness in question, Jack felt the heat. But he and his team still saw the same potential as always, even if the payoff was presently delayed. The problem was in the human piece, not the technical system. Jack and his team had underestimated the effort required to gain widespread buy-in and compliance. It was time to follow through in that effort.

So they doubled down on change-management efforts. At each shipping center, the training team would have to stick it out, refusing to leave until performance results were attained. Transferring knowledge wasn't enough. The center's staff might be fascinated by the new system, but that excitement was often just a flash in the pan. Left on their own too quickly, they would return to old routines.

How do you reform stubborn creatures of habit? There's always sheer will and an iron fist. Big change requires some law enforcement. The team supervised, cajoled, and even micromanaged a bit. For example, loaders who struggled to break old habits were reassigned to new areas with which they weren't familiar, where they wouldn't recognize delivery addresses. You can't take the knowledge out of a person, but you can take the person out of their domain of knowledge.

But babysitting and arm-twisting go only so far. Rather than relying solely on the application of pressure, Jack's team mobilized by sharing the rewards of success. The trick was to reward in terms of short-term success since improvements to bottom-line efficiency would take some time to materialize. "Because those early transition days are not necessarily profitable, we had to use a *balanced scorecard* that would reward managers who achieved leading indicators," Jack explains. "If you're doing these leading things that are in your control, how can the lagging indicators of dollars saved not follow?"

The team implemented scorecards that reported on staff adherence to the improved procedures, flagging when there were more than a small number of overrides or when drivers would have to wait for their truck to finish loading and depart late. Only after a passing grade would the shipping center "graduate" and the training team leave.

This performance-management tactic worked. It increased adherence to centralized decisions and decreased the number of decentralized decisions made on the fly. Off came the training wheels.

Achieving these quicker, incremental wins changed the conversation, gaining renewed support from the top. The budget and available resources nudged up and the training team grew to cover more shipping centers. A typical shipping center required five training personnel working on-site for many weeks. To meet the extraordinary demands of this full-scale change-management process, Jack's deployment team ultimately grew to about 450 (and later to 700 for the ORION navigation system described in the introduction).

But as positive results began to show, a new concern came into focus: How could Jack prove that the improvements were from his PFT optimization system rather than from other changes? Across UPS, a plethora of overlapping efforts were in play, all attempting to improve operational efficiency. The onus was on Jack to somehow demonstrate that his system was the cause of most of the observed improvements.

When You Don't Need Humans in the Loop

With the EduPay project, I faced the same two deployment challenges as Jack: (1) getting models properly integrated to take effect and (2) establishing due credit for the value they generated. He was delivering packages and I was delivering ads, but ML's challenges are universal.

Fortunately for me, the first challenge, integration, was more straightforward to address for my project with EduPay. For one thing, the scale was about one one-hundredth UPS's scale. Every day, his system would have to decide how to route each of 16 million packages. But mine was still hefty, deciding which ad to show about 200,000 times a day. The smaller magnitude did little to alleviate the engineering challenge.

More than the difference in scale, the main way I had it easier was that my project would automate decisions rather than support human decisions. I got to knock the human out of the loop. That makes for a simpler project. It's easier to get computers to follow instructions. After all, that's what they're built to do. Jack needed hundreds of people to train the humans. I only needed an engineer or two to reprogram an aspect of EduPay's website.

ML projects often automate decisions, such as with response modeling, churn modeling, and spam filtering. Each time a model's output determines whether to contact a customer or relegate an email message to your spam folder, there's no human in the loop. The system acts autonomously.

>**Decision automation**: The deployment of a predictive model to drive a series of operational decisions automatically.

When humans are out of the loop, they aren't out of the picture. Automation streamlines a central piece of the process, but humans are still involved somewhere down the line. After a model targets a marketing campaign, even if you manually lick each stamp, it's still *decision automation*, since the batch of yes-contact/no-contact decisions was made unilaterally by the model. For some fraud detection projects, the model decides which transactions to manually audit for fraud, but this decision is automatic, and, indeed, most transactions go through without human involvement. A credit score may determine that some small loan applications go through automatically, some are denied automatically, and some are sent to human loan officers to render a final decision. Since the assignment to those three bins is automatic for each application, some would call this "decision automation," but others would call it "partial automation."

On the other hand, sheer complexity prohibited the UPS project from implementing decision automation. The project involved an unusually long distance between prediction and decision. Deliveries are predicted and these inform a sophisticated planning system that assigns both predicted and known deliveries to trucks so that each truck's route that day will be efficient. The predictions affect the final decisions, but more indirectly. Due to operational complexity, the decisions are only semi-automatic, with human experts modifying them as needed.

But plenty of more straightforward deployments also involve humans in the loop. Even when each prediction informs each decision in a direct, clear-cut manner, full automation isn't usually a consideration for human resources, healthcare, and law enforcement, for example. In those arenas, computers don't have the final word. Only humans can make the weighty decisions of whom to hire, how to diagnose or treat a patient, or whether to parole an inmate.

By supporting human decisions, a model augments rather than automates. For example, if a model indicates that a job applicant is very likely to succeed, this may affect a hiring manager's thinking. Or a model may signal to a customer service representative that the individual to whom they're speaking is likely to cancel their subscription. The representative can take this under advisement as they like, depending on how the conversation is going. By working together, human/machine teams often prove to outperform either party alone. As American economist Leo Cherne put it, "The computer is incredibly fast, accurate, and stupid. Man is unbelievably slow, inaccurate, and brilliant. The marriage of the two is a force beyond calculation."

> **Decision support**: The deployment of a predictive model to inform operational decisions made by a person. In their decision-making process, the person informally integrates or considers the model's predictive scores in whatever ad hoc manner they see fit. Also known as *human-in-the-loop*.

In general, companies prefer decision automation when feasible. Often, the bottom line is served by leaving humans out of the loop so we can leverage the sheer speed and efficiency of the machine. In that case, we dispense with much of the effort needed to train people—but we must provide the machine with a bit of extra instruction that supplements the model predictions themselves so that they can be acted upon.

Translating Predictions to Actions

For EduPay, predictions drove ad selection in a relatively direct manner—but you always need some customized logic to top things off. A user arrives on a web page, so it's time to pick the ad. The system has 291 models—one per ad—and uses them to derive 291 probabilities. Each one tells you the chances that this user would respond if shown the corresponding ad. If you choose the ad with the highest probability, you're most likely to receive a response from the user.

But there was another factor at play: each ad's bid, that is, the amount the sponsor would pay for each response. No surprise: EduPay's aim was to increase revenue, not clicks. So the *expected return* for each ad was calculated, simply by multiplying the probability of a response by the bid. For example, consider choosing between these two ads:

Ad A has a probability of 20 percent and a bid of $5, so the expected return is $1.

Ad B has a probability of 10 percent and a bid of $15, so the expected return is $1.50.

The system will choose ad B, even though it has a lower probability of response. By running things this way, we don't maximize responses, but we do maximize revenue. Recall the chapter 3 sidebar where we similarly transformed the probability that a debtor would repay into the expected return by factoring in the revenue generated by a loan.

Beyond this, EduPay also had to filter ads by eligibility. Sponsors would only pay for responses by eligible candidates, such as certain military recruitment pertaining only to users who were seventeen years old or older.

In the end, simple logic and arithmetic translate predictions to actions. For each ML project, leaders manually design this translation based on business pragmatics and requirements. The translation scheme may be detailed and particular, but for most ML projects, it's not terribly complex.

For card fraud detection, model deployment and score translation are carried out separately, by two different companies. FICO deploys the Falcon model for a bank. For each transaction, it delivers a score between 1 and 999—a cosmetic spin on the traditional 0-to-1 or 0-to-100 range within which probabilities are normally expressed, which FICO employs only to make the scores feel more friendly to the bank. It's then entirely up to the bank how to act on the scores, depending largely on their tolerance for fraud in comparison to their tolerance for interrupting customer purchases. For example, a bank could implement rules that treat different risk levels differently, depending on the dollar amount of the transaction, such as:

> If the charge is more than $500 and the score is more than 950, then decline the transaction.
>
> If the charge is more than $100 and the score is more than 980, then decline the transaction.
>
> If the charge is more than $100 and the score is more than 900, then contact the customer to confirm the charge.
>
> . . .

Such rules usually fall outside the jurisdiction of data scientists. Banks develop them manually, based on policies, regulations, and business strategy. The rules are not generated automatically—they embody how the company has decided to use a model, rather than being part of the model itself. Small banks may employ somewhere between fifty and 200 such rules, while a large bank could have thousands. In the end, depending on the rules set by a bank, the system usually winds up intervening about fifteen to thirty times for every 1,000 transactions.

There's no secret sauce required to translate from prediction to action. Even while the logic may grow in detail, this translation doesn't require any form of advanced analytics. Beware the often-invoked misnomer *prescriptive analytics*, which falsely implies that you need another kind of sophisticated technology beyond predictive analytics to move from predicting outcomes to prescribing actions. Predictive analytics, the use of ML for certain business applications, is already intrinsically prescriptive. Its purpose is to prescribe actions and it alone already gets you quite close to doing so. For the last few inches, you only need carefully crafted, customized logic—not a whole new class of analytical methods. Needlessly introducing the term "prescriptive analytics" has caused confusion by implying the presence of novel advanced methods where none exist.

With an established method to translate model scores to ad selection, the EduPay project was almost ready to launch. Next, we had to embed the models into a live website.

How to Export a Model

EduPay engineers were at the ready to integrate my models—but they were entirely new to ML. What does it take for a quant to transfer his model to a group of engineers?

To deploy a model, you must set it free. You've got to export it out of the ML software tool and into the operational system. Nowhere is this send-off more literal than when you embed a model within a mobile device. Within each iPhone sits a neural network for face detection. Meanwhile, the Google Pixel 6 was the first phone to house a separate chip to run models.

But even when you aren't deploying to an untethered device, your model must migrate just the same—to the hands of coders who will integrate it as the functional apparatus it was meant to be. Most ML software tools facilitate this in a couple ways. One is to generate code. At the press of a button, it translates the model's mechanics into Python or C so that an engineer can more easily integrate it into an existing system. Alternatively, an engineer can create a stand-alone software module that houses the model and is invoked via an *application programming interface (API)*—which is a standard mechanism by which one system can call on another. This provides to the engineers a function that they can invoke from within their code, passing to it the details about an individual—that is, the model inputs—and receiving in return the predictive score for that individual.

I didn't have it so easy. Since I'd programmed the modeling method from scratch, I had no industrialized tool. To make matters worse, I wasn't an engineer. Although I'd been programming since I was ten, I'd never pursued a career in it. As an academic-turned-analytics consultant, I knew what needed to be done conceptually, but I was in no position to personally translate my models into production-ready code and integrate that code into an operational system.

So, I spelled it out for them. In a dense, three-page, single-spaced Word document entitled "Scoring Module Requirements," I wrote down every single detail the engineers needed, every step they would need to code in order to use the models to predict responses to ads. For a given user and a given ad, it described how their code should look up a bunch of values within a table I'd created—the table had about 50,000 rows of values that constituted the 291 models—and then apply the right arithmetic. The document ended like this:

> Finally, after calculating the relative probabilities of a response (p1) and of no response (p0), normalize the two with the formula p1/(p0 + p1). This is the absolute probability of a response, that is, the score output by the model.

Given the characteristics of an individual that make up the inputs to a model, calculating the score isn't all that complex—especially with the simple kind of model I'd used for EduPay, Naive Bayes. But pulling together those inputs on the fly is a whole 'nother ball game.

The Data Disconnect: Getting the Inputs to a Deployed Model

> Getting the data right and having it in the right place at the right time is 80–90 percent of the problem.
>
> —Scott Zoldi, chief analytics officer, FICO

If the struggle to deploy predictive models is a battle, then the challenge of hooking up its inputs is right at the frontlines. Somehow, a deployed model must receive the right set of values each time the model is invoked. At the moment a model is to score an individual case, it needs its inputs—the values that characterize that case. Having those inputs at the right place at the right time may be the very trickiest engineering challenge when architecting for deployment.

The problem stems from the *data disconnect*, an abominable divide between model development and deployment. When preparing the training data, the data scientist is typically focused only on incubating a model and ensuring that it performs well in "the lab." To that end, they set up the input variables—positioned as columns in the training data—in whatever ad hoc manner is most convenient.

This leaves a formidable challenge for deployment. The system housing the model will need to recreate the variables exactly as the data scientist set them up during development, mimicking the form and format they held within the data scientist's system or within the ML software, both of which are typically foreign to the engineers.

In that endeavor, every detail matters. For example, my EduPay models took the email domain as input. Should it be a string of characters like "yahoo" or "gmail"? Or should it also include the ".com"? Must it be all lowercase? Should Boolean variables like "US citizen—yes or no" or "Has opted in for marketing email—yes or no" be represented as 1 and 0, "yes" and "no," or "Y" and "N"? How do you represent a value that's simply unknown, a.k.a. a *missing value*—is it the word "NULL," an empty string, a negative one (–1), or something else? How do you calculate the SAT verbal-to-math ratio if the math score is unknown, considering that dividing by zero is impossible?

When it comes to transferring a model from one system to another, it's like we're stuck in 1980 typing commands at a DOS prompt with no spell check. Get any detail wrong, and the model doesn't work as it should.

To make matters worse, model inputs may originate from various siloed sources across the organization. Since the inputs were designed to comprehensively represent much of what's known about an individual, the databases that hold them could reside across disparate systems. For example, demographics may come from a customer relationship management database, while variables such as "Already seen this ad before—yes or no" may only be available by scanning an operational log to check. Pulling these together on the fly at scale during deployment presents an engineering challenge that data scientists often fail to anticipate.

It's a tough job. According to a 2021 survey of data engineers, 97 percent feel "burned out" and 78 percent wish their job came with a therapist. Although that's not a joke, the report, by DataKitchen and data.world, couldn't resist asking, "Tell me about your motherboard."

One Firm's Firm Approach to the Data Disconnect

The antidote to the data disconnect? A new connection. Model development and deployment must be bound and inseparable. The two have traditionally been handled discretely, as isolated steps—conceptually linked yet decoupled in practice—but successful leaders seek to unify them so that preparing the data for modeling and engineering the inputs for deployment are one and the same.

But this means asking data scientists to change their habits and to accept some new responsibility. Many have grown accustomed to thinking up and implementing input variables at will during the model training step—without paying heed to how they'll be made available during deployment. With a focus on developing and evaluating models offline, they view engineering as a distinct job, department, and mindset. Data scientists often see themselves in the business of prototyping, not production.

Nothing breaks techie habits like executive authority. Enter Gerhard Pilcher, the president and CEO of Elder Research, a widely experienced data consulting firm with which I've collaborated many times. Gerhard has instilled best practices across the firm's client projects that have data scientists collaborating in detail with data engineers from the beginning of each modeling effort.

I asked Gerhard if he had implemented this change with a rule prohibiting data scientists from cobbling together their training data in a vacuum. He shied away from "rule," but he put it this way: "We discourage ad hoc data aggregation. That change took a little while to take root." His firm but friendly leadership ushered the team through a culture shift and into a new paradigm.

Under the guidance of this improved practice, data scientists request the model inputs that they will want available for model deployment from the data engineers rather than only hacking them together on their own for the training data. It's a bit less impulsive and a bit more team spirited. With this process in place, the data infrastructure to support deployment—called the *data pipeline*—is already being constructed even during the model training step. Come deployment time, the process to deliver inputs on the fly is repeatable and reliable. This is because the pertinent data sources have been pre-connected during model development. This way, "once you've tuned and validated the model," Gerhard says, "you can deliver the result much more easily."

By designing the data pipeline early, you not only proactively prepare for deployment—you also win by recognizing infeasibilities early, moving up project decision points and even failing fast when needed. Since some data sources can be costly to integrate, "the client will experience sticker shock," warns Gerhard. "We can preempt that shock and ease the blow, or cancel if necessary. The sooner you kill an effort that's not deployable, the better."

This makes deploying ML projects a scalable endeavor. My EduPay project would have benefited—without it, I had to brute-force my way to deployment by painfully detailing the inputs' calculations within my "Scoring Module Requirements" document and hoping the engineers would get all of it right. FICO Falcon, on the other hand, has by now streamlined its data pipelines by sheer repetition, since each time FICO sets up the system for a new bank, the same inputs must be calculated with that bank's data.

Beyond the data disconnect, Elder Research has also learned other hard lessons about the change-management challenges of deployment, the struggle to gain acceptance from those on the ground—much the same as the lessons UPS has learned. ML "often dictates a major change in how people act," says founder John Elder. "Many people revert to the old way of doing things instead of trusting the model. We studied this and found several ways to improve the environment of trust—both technical and interpersonal. People (often rationally) fear change. They don't want to abandon the way they make decisions. The most important way to address that is to work side-by-side with potential allies from the very beginning and earn their trust."

These process improvements worked. By implementing them, Elder Research boosted its deployment track record. During the first decade after the company was founded in the mid-1990s, only 65 percent of the models they developed for clients were deployed, even though 90 percent met predictive performance requirements. This success rate was about three times higher than that of the industry as a whole, but the firm was determined to do better. By implementing these new practices, over the following ten-year period, the firm's model-deployment rate soared from 65 to 92 percent, and its model performance success rate rose from 90 to 98 percent.

The proactive tactic of establishing a tight connection between model development and deployment is a key ingredient for success. But our work is not over yet. After resolving the data disconnect, one major engineering challenge remains: ensuring the model operates quickly enough.

The Need for Speed: Driving Decisions in Real Time

For some deployments, models must act fast. When an EduPay web page loads, the chosen ad must appear immediately. To that end, the models must drive their decisions instantaneously. When a fraudster attempts to perform a card transaction, FICO Falcon must act quickly enough to block it. Autonomous vehicles must recognize obstacles quickly enough to steer or brake.

For other deployments, unhurried, offline scoring fits the bill. For example, take direct mail targeting. You may have 10 million contacts, each to be scored according to the likelihood they'll buy if you send them a brochure. But even with that many, your system would only need to score a few hundred per second to complete the task overnight, or just a couple dozen per second if you have a few days and a few computers.

The same is true for other kinds of operations. Offline, batch processing often suffices for purchase orders, insurance claims, banking checks, or applications for insurance coverage or lines of credit. For these uses, introducing scoring to an existing process doesn't usually impose intense performance challenges. Higher speeds could still improve organizational efficiency, but we don't need to scrutinize each millisecond.

For these applications, you don't necessarily even need to export the model. Many ML software tools can apply a model to score a batch of cases from within the tool itself. You simply point the tool toward a data table of new individuals, and it generates the score for each one, which can be tacked on as a new column of data. This use of a model is still called *model deployment* even though the model itself isn't exported.

But many of the greatest business opportunities for ML require real-time scoring. This is because high velocity means high volume—the operations that take place most quickly happen most abundantly. To optimize these largest-scale processes, model scoring must take place in real time, at the moment of each interaction.

On the web, speed is of the essence. According to Google, when search is half a second slower, traffic and revenue suffer by 20 percent. Similarly, in its experiments, Amazon showed that even web-page slowdowns of 100 milliseconds result in a "substantial and costly" drop in revenue. Booking.com found that an increase of 30 percent in latency costs about 0.5 percent in conversion rates—"a relevant cost for our business"—and the web analytics firm Kissmetrics reports, "A one-second delay in page response can result in a 7 percent reduction in conversions."

Latency also clobbers automated trading since a handful of milliseconds can mean a missed price opportunity. Estimates show that if an electronic trading system lags 5 milliseconds behind a competitor, this could cost $4 million per millisecond.

Speedy Delivery: Models Work Fast

Great news: Model scoring can be fast, as fast as most any project needs it to be. Model scoring is not the *learning* part of machine learning—that's the "heavy lifting." Rather, it's the application of what's been learned. To score with a model is usually only a matter of applying a fixed mathematical formula that does not involve any loops. Computers do so super quickly.

The previous step, model training, consumes the most time. It must operate across an entire set of training data, which could consist of hundreds of thousands or millions of learning cases. That algorithm eats up many computational cycles to generate the model, running through the data with its trial-and-error process. Model training is typically executed as an offline process that doesn't need to utilize real-time systems. In rare cases, an already-deployed model is continuously updated as new training examples are encountered, but such *online learning* is very uncommon, since it's complex and costly to implement and it fails to deliver—if properly scheduled, periodic batch training for refreshing a model usually results in model performance that holds up virtually as well.

Once trained, models work fast. To calculate the score for an individual, models need only operate on the input data for that one individual. And since the model itself is often a relatively simple structure, applying it can be a relatively lightweight step for the machine. For example, a logistic model is simply a weighted sum of the inputs, with a bit of nonlinear "squeezing and stretching" added on for good measure.

Of course, the right computer hardware must be in place to score in real time. But you may well already have it: the existing hardware that's currently running the high-speed operations you're aiming to improve. For many projects, high-performance systems that have already been optimized for online functions can potentially also handle the relatively light additional task of scoring, incorporating it so that there's only a miniscule impact on speed.

For detecting payment card fraud, models often deploy on a mainframe computer. Often misunderstood to be legacy technology, mainframes have simply never stopped advancing over the decades. They achieve very high velocities and are so reliable that they typically run for more than a decade without any outages. Financial institutions use them to process 90 percent of all credit card transactions.

Depending on the bank that's using it, FICO Falcon sometimes deploys its fraud model on a mainframe and sometimes on less expensive systems such as a Linux server in the cloud. Either way, the aim is to score each transaction in less than 30 milliseconds, including the latency incurred if each request for a score must travel to the cloud and back. The company has confirmed that many banks take less than 100 milliseconds to score each transaction, sometimes averaging only 10 milliseconds.

Mainframe manufacturers are quick to point out how much cloud deployment can slow things down. Jonathan Sloan, who runs marketing for ML solutions that run on IBM's Z system, which is the frontrunner of the mainframe market, conducted experiments to compare the speed of scoring with an on-premises mainframe to scoring with calls to a computer in the cloud. His results showed that the cloud can multiply the time it takes by a factor of over 80, for example, from 1 millisecond to 80+ milliseconds. With a mainframe, "Organizations can achieve much greater throughput, dramatically better response times, and greater confidence in meeting service-level agreements," he and a colleague wrote in a white paper.

Either way, whatever kind of system you already have handling large-scale operations, it can usually also handle the model. A bank that's processing hundreds of thousands of card transactions a day can introduce one more step for each one: scoring with a fraud detection model. Likewise, EduPay's web servers were already hosting its heavily visited website and they could also handle model-scoring in order to improve ad selection each time the site served an ad to a user.

Organizations *can* deploy in real time—we have the technology—but that doesn't mean that they *will*. Many still hesitate.

The Greatest Opportunities Are the Hardest to Tap

Organizations often fumble the greatest opportunity that ML has to offer: optimizing the largest-scale operations. Since they're often the highest-speed operations, they require real-time deployment, which is harder to greenlight. It's more complex and carries greater risk, since it means changing mission-critical, high-speed systems.

As a result, integrating real-time predictive scoring is more rare and cutting-edge than many realize. During its formative decades when the ML industry was incubating and developing—and building a reputation—most models were deployed not in real time but only offline in "batch mode," for applications that didn't require real-time scoring, such as targeting direct marketing and scoring credit applicants. For most ML projects, this is still the case.

Deploying in real time increases the potential gains—but also the resistance to change. It's only human nature. The larger the scale, the greater the fear. If you suggest a plan to enhance operations that are currently flying by at thousands of transactions per second, some of your colleagues just might freeze up. Some will argue that the company can't possibly afford the cost of introducing a new step to each and every transaction—nor the risk that doing so could slow things down.

But neglecting to move forward incurs a severe opportunity cost and puts the organization's competitive stronghold in jeopardy.

Back to basics. The six-step practice of bizML overcomes resistance by making known what is otherwise a fearful unknown. Only by engaging and ramping up decision makers on the end-to-end plan—so that they fully understand both the value and the technical feasibility of deployment—can a leader overcome an otherwise costly case of analysis paralysis.

One simple clarification especially greases the wheels for the green light: Runtime operations remain unaffected by the heavy lifting of model training. Data scientists train the model somewhere else, sparing online systems from that burden so that they may continue running smoothly. Those well immersed in ML may neglect to make this abundantly clear, but this distinction will assuage stakeholders, who are understandably protective of operational systems. They will be relieved to hear that there's no need to strain real-time systems that are already handling operations, that the heavy lifting carried out by modeling algorithms is kept apart, handled by separate resources allocated to data scientists for model development.

I was pleased to find that EduPay was both capable of and confident about deploying my models—they could and they would. But confidence is never literally 100 percent. Rather than swapping out the existing ad-selection system entirely, it would be safer to start with an incremental step.

Mitigating Deployment Risk with a Control Group

Melissa, the director at EduPay who'd brought me on board, proposed a prudent next step: Deployment would start by using the models only half the time. For half the users, nothing would change, but the other half would see ads selected by the models.

This is *A/B testing*, but not the way web marketers usually conceive of it. Typically, you set up a head-to-head comparison between two simple options. Color 1 versus color 2. Product A versus product B. But in this case the website would compare a complex ad-selection method with an even more complex, model-based ad-selection method. Data scientists call this a *controlled experiment*.

Whatever you call it, this tactic measures how much the new method improves the business metrics over the existing legacy method by trying them both simultaneously. This is critical since, if you try one for a predetermined time period, stop, and then try the other, then you are not *controlling* for other changes that may have taken place in the meantime. You can't directly control all sorts of factors that might make the comparison unfair, such as seasonal shifts, changes in customer trends, or other concurrent operational changes. As a result, even if performance improves with deployment, you can only be sure it resulted from the model (or models) if you simultaneously track a *control group*.

Beyond establishing credit where it is due, a control group serves another critical need: mitigating deployment risk. Many things can go wrong when you deploy, resulting in performance that is disappointing if not disastrous. After all, any new system might have bugs—and in the case of model deployment, you've opened yourself up not only to logical or programming bugs but also to quantitative bugs, including mistakes in the math or misconceptions about the data.

Fortunately, you can manage this risk as conservatively as you'd like by deploying incrementally, just a bit at a time. For example, rather than jumping to Melissa's 50/50 champion/challenger runoff, you could start with an even smaller, more incremental step by introducing the model-based process only 5 percent of the time. After this, as trust increases, grow from there. Along these lines, UPS also started with only a partial deployment, initially integrating the system at only a few shipping centers.

Without a control group, it's all too easy to grant undeserved credit to a model. This is a common mistake that's often made, for example, with direct marketing. If a targeted campaign elicits a high response rate, stop and think for a moment before you congratulate the marketing manager. It may be that the model is doing a great job identifying customers more likely to buy, but that most of them would have bought anyway, even if not contacted. In that case, the money spent mailing brochures may not have been making an actual impact on sales. A control group would avert this pitfall by righting the misconception. If you simultaneously observe the purchases made by a control group of customers with whom no contact was made, then you have your baseline for comparison.

For EduPay, the control group provided a critical gauge, since we were moving to deployment with a lot of uncertainty. We had no solid method for estimating the gain beforehand. We could evaluate individual models, but the combined effect of using all 291 models for each ad choice was an unknown before deployment. Improving over the existing legacy method was not a sure thing since it was a tough champion to unseat. Even though the existing method didn't personalize ad choices based on the particulars of each user, the fact is that you do well by simply serving up the most universally popular, high-paying ads that a user hasn't yet seen. That's a standard approach for online ads. Any time a model-based system goes with a less universally popular ad, it takes a risk.

It's no wonder that Gary Loveman, when he was at Harrah's casino—a CEO with a PhD in economics from MIT—famously said that he'd fire any employee who runs an experiment without a control group just as quickly as one who steals from the company.

After this champion-challenger runoff, the EduPay results came in. In comparison to the control group, model-based ads boosted revenue by 3.6 percent, enough to aggregate an extra $1 million every nineteen months—and possibly more, if extended from *interstitial* ads to also target those embedded within other web pages. It achieved this improvement by sometimes selecting ads that were less universally profitable, but that the individual customer was more likely to click on. As a result, the overall response rate increased more than revenue did—by 25 percent. It's safe to assume that this higher rate of response meant that the users were now experiencing even greater ad relevancy, more often seeing ads that served their interests.

When it comes to large-scale systems, a boost of a few points goes a long way. According to McKinsey, "Our research finds that for each $5 billion in credit balances a bank originates, an increase of just one percentage point in the predictive power of a credit model could reduce losses by up to $10 million within the first year alone."

Credit Where Credit Is Due: A Control Group at UPS

UPS also saw the same potential from a small win. As Jack at UPS put it, "Little things matter. If we can just reduce one mile per driver per day in the U.S. alone, we can impact the bottom line by $50 million . . . one minute per driver per day is worth $14.6 million."

Big or small, Jack needed a win. Having doubled down on deployment training with large teams devoted to the task, there was every reason to believe business performance would benefit.

And Jack had a control group against which to compare performance improvements: all the UPS shipping centers that hadn't yet deployed his PFT system. Since deployment had started at only a limited number of trial locations—a sample that was considered representative of shipping centers in general—the performance at all the other sites served as a control.

Initial signs were good. Within a few months of deployment, it became clear that performance had improved by 15 percent over control sites, according to a highly visible key performance indicator (KPI) within UPS: *stops-per-mile*. The more efficiently a truck's route was planned and utilized, the more delivery stops it would make for every mile it drove. As stops-per-mile increased, the aggregate miles, gas, and driver time needed to fulfill a day's deliveries decreased.

PFT flourished, eventually gaining notoriety as a success: It was saving 85 million miles annually. The press was ready to congratulate rather than eviscerate. *InformationWeek* even placed the project atop its annual "20 Great Ideas to Steal" list.

This great gain came from the deployment of the package-prediction model in combination with other related improvements, such as centralizing the package-delivery planning at each shipping center. Jack informally credits the predictive model itself with an estimated 10–25 percent of PFT's wins, although it is hard to differentiate between the contributions of mutually interdependent innovations.

The rest is history, as already told in this book's introduction: Jack built ORION on top of PFT and convinced the executive Chuck to authorize the system, which prescribes turn-by-turn driving routes. The overall efficiency compounded even further, ultimately saving the company 185 million miles and 185,000 metric tons of emissions every year.

The End Is a New Beginning

"Happy ending" is an oxymoron—if a good thing ends, you're *un*happy. Once you deploy a model, you've only just begun to reap the benefits. Likewise, you've also only just begun to maintain the model. To keep it in play and sustain its effectiveness, you must monitor and periodically refresh it. Proceed to this book's conclusion for the lowdown on model upkeep, as well as a few other practicalities for your ML project: how to sell it, who to enlist for it, how long it takes, and how to responsibly manage its societal impact.

BizML Cheat Sheet

The Strategic Playbook for Machine Learning Deployment

1. Establish the deployment goal (value)

Define the business value proposition: how ML will affect operations in order to improve them.

2. Establish the prediction goal (target)

Define what the ML model will predict for each individual case.

3. Establish the evaluation metrics (performance)

Determine the salient benchmarks to track during both model training and model deployment and determine what performance level must be achieved for the project to be considered a success.

4. Prepare the data (fuel)

Define what the training data must look like and get it into that form.

5. Train the model (algorithm)

Generate a predictive model from the data.

6. Deploy the model (launch)

Use the model to render predictive scores and then act on those scores to improve business operations.

After step 6: Maintain the model (upkeep)

Monitor and periodically refresh the model as an ongoing process.

Key Execution Strategy

All steps require deep collaboration with business stakeholders.

Business stakeholders must hold a semi-technical understanding of ML.

The steps are not executed linearly—backtracking prevails.

—from *The AI Playbook* by Eric Siegel

Conclusion

ML's Elevator Pitch, Staff, Timeline, Upkeep, and Ethics

Now that we've seen the entire end-to-end process, it's time to distill it back down to a brief proposition. This book concludes by describing how a machine learning initiative must begin: pitching the project. In a nutshell, you must sell the way in which ML will launch and the value of doing so. This conclusion also rounds out the bizML practice with the who, how long, and then-what: who constitutes the project team, how long it takes, and then what ongoing maintenance you must perform to keep the model in operation. Finally, I end with the enormous ethical responsibilities you accept when you deploy ML.

When you're selling machine learning deployment, swimming upstream against resistance and inertia, it sometimes feels like you're hustling. But in actuality, you're recruiting. You're enlisting collaborators and orchestrating a vision. Don't get me wrong; as you advocate for the project, there are times you might need to aggressively cajole. During parts of Jack Levis's story, it looked like a battle of wills between him and a universe of naysayers. But sometimes it takes a hard sell—not only to prevail and achieve buy-in, but to catalyze a fruitful, unified collaboration across the enterprise.

At first, the imbalance is real: You get it and they don't. And they may be slow to come around. But once they finally do, it's up to you to embrace a profound shift in perspective: More than a skeptic you've convinced, they're now a critical partner who will deliver new insights that the project needs. Only by having sold them on your plan can you now jointly refine and improve it.

When you first pitch, bending over backward is part of the deal. You may feel like this stuff should basically just sell itself. After all, the value proposition can seem totally obvious when you're already invested in it. The potential operational improvement is a "no-brainer." But, to get the green light, you must get the people in charge not only interested but enthusiastic. This means taking a step back from the excitement and telling a simple, non-technical story that is dispassionate rather than fervent, one that could just as well come from the lips of a truly impartial third party. In the art of sales, evenhandedness is more rousing.

To sell ML convincingly, sell it succinctly. Strengthen your pitch by distilling it down to the fundamentals: the precise operational change, the value of that change, and how ML will achieve the change—*in that order*. It's time to perfect your elevator pitch.

The Elevator Pitch

> No one wants to be sold to, but everybody loves to buy. Give them something to buy.
>
> —Jack Levis

The premise to this book's bizML practice is simple: Reframe "ML projects" as "operations-improvement projects that use ML." Leading with the scientific virtues and quantitative capabilities of the technology—such as modeling algorithms, the idea of learning from data, or the notion of probabilities—is putting the cart before the horse. Instead, lead with the business value proposition, a simple story about how processes will improve.

Here's an example elevator pitch:

> Currently, 99.5 percent of our direct mail is ineffective. Only half a percent respond.
>
> If we could increase that to 1.5 percent, that would mean a projected $500,000 increase in revenue in return for our current

marketing spend, tripling the ROI of marketing campaigns. I can show you the arithmetic in detail.

ML can hone down the population to whom we're marketing by targeting the customers more likely to respond. This should deliver the gains and ROI I just mentioned.

What do you think? Would you support this project or would you have objections? What questions do you have?

When you pitch, get straight to the point, the business value and the bottom line, and then gauge the person to whom you're speaking. They will be interested in the business value, but they're not necessarily excited about ML. ML is only the technical solution, the means to the end, so, in this early stage, its details can easily distract, confuse, or bore.

Your narrowly focused pitch must accomplish three things:

1. *Lead with the value proposition*, expressed in business terms, without details about ML, models, or data. For now, share nothing about how ML works, only the actionable value it delivers, the operational improvement gained by model deployment. This usually means avoiding the words "model" and "deployment."

2. *Estimate the value*, a performance improvement in terms of one or two key performance indicators (KPIs), such as response rate, profit, ROI, cost reduction, or labor reduction. You must include a potential KPI win, even if only from scratch calculations. Convey this potential in simple terms, such as a bar graph that has only two bars to illustrate the potential improvement. It's not yet time to mention predictive performance metrics such as lift. Make the case that the KPI win will more than justify the expense of the ML project.

3. *Stop and listen.* Keep the pitch short and then open the conversation. Realize your pitch isn't the conclusion but rather a catalyst to begin a dialogue. By laying out the fundamental proposition and asking them to go next, you get to find out which aspects are of concern and which are of interest, and you get a read on their comfort level with ML or with analytics in general.

After the pitch, you've got to interactively gauge when to get into details about how ML will be applied—and at what depth and speed. It's more common than you may realize for the business professional to whom you're speaking to feel nervous about their own ability to understand analytical methods. People are skilled at covering this nervousness.

Keep it simple. As with many technologies, convolution and the appearance of arcane complexity threaten to extinguish a newcomer's excitement about the potential value. This might leave them feeling compelled only by the pressure that comes from all the "Everyone's doing it!" hype. Nip that in the bud with a straightforward, concrete explanation. Cover just enough of the inside mechanics to demystify ML.

Resist the temptation to ride the wave of "AI" hype. It oversells. The propaganda's sheer excitement does successfully broadcast that there's value to be had—but it only distracts from the concrete value proposition by idealizing the core technology. Don't passively affirm starry-eyed decision makers who appear to be bowing at the altar of an all-capable AI. If you do, here's the risk that you face: When the hype fades and the overselling is debunked, much of ML's true value proposition will inevitably be disposed of along with the myths, like the baby with the bathwater.

Exercise Patience and Solicit Input

One of the hardest things I have to teach my employees in the art of consulting is you have to talk to people much more often than you want to.

—John Elder, founder, Elder Research

Data scientists will literally solve AGI instead of talking to a product manager.

—Josh Wills, experienced data scientist

As decision makers slowly come around, you may find that they do so by way of a long, winding road. The first meeting is only one of many. Ramping up others on the proposed initiative is, well, a process.

In the end, it's you versus the Fear of Change. Volumes have been written on change management. In fact, that's the theme of this book as well. It covers the practice and the background knowledge for managing the particular kind of change made by ML deployment.

But no preordained plan of action is bulletproof against the wild card you will ultimately face: human anxiety. "We can't afford the cost of integrating a model into mission-critical operations," nervous executives will declare, "nor can we afford the associated risk of doing so."

Your response is simple: We can't afford not to. Turn up the heat from there as you see fit. If we don't jump on the opportunity, one of a diminishing number of unique differentiators that technology can provide, we will incur a severe opportunity cost and jeopardize our competitive stronghold. Streamlining operations with models is not a question of whether but of when: before the competition does or after? Change can be hard, but the facts are much harder. If pure apprehension is precluding the company from pursuing the value propositions that ML has to offer, then the business is getting in the way of doing business.

Even as you're selling, you're learning. Be prepared to change course. You can't know beforehand what objections, valid feedback, and new information may arise. You're there to listen as much as you are to talk. Undoubtedly, you will learn about new pragmatic considerations that mean modifying the operational deployment you've had in mind.

Ultimately, your persistence will pay off—but only after providing more information than you've crammed into the elevator pitch. Management needs more specifics, including the staffing requirements and project timeline, in order to finalize their decision. And you will need these specifics too, in order to properly execute. Let's dive into them.

Assemble Your Team: Staffing the ML Project

At a minimum, in addition to the project lead—sometimes called the *data product manager*—you need technical experts to facilitate each of the three culminating project steps. Be warned that there's not a lot of agreement on terminology for these roles. Here's the breakdown:

For step 4—prepare the data—you need a *data engineer* or *data wrangler*, someone familiar with the data tables in their current form and capable of transforming them into training data. This person is responsible for accessing and reconfiguring the data. This task will often be split across multiple people, since it involves miscellaneous tasks often assigned to certain database administrators and database programmers, and it involves multiple technologies such as cloud computing and high-bandwidth data pipelines.

For step 5—train the model—you need a *predictive modeler*, a hands-on expert in core ML methods. This person creates the model using ML software that operates on the training data. Often, this person takes on the more general title of *data scientist*, so that's the title named most often throughout the chapters of this book.

For step 6—deploy the model—you need an *ML engineer* capable of modifying the existing operational system so that a model is newly integrated. The engineering requirements will vary greatly depending on what kind of operational change you're making and how the operational system was constructed in the first place.

For a pilot project—an isolated, narrowly focused ML initiative—you may not need more than the skeleton staff listed above. With such a project, you start judiciously small, and so your staffing investment should start small as well. On the other hand, if your project is part of a broader analytics initiative that will spawn multiple projects across the enterprise, that's a very different story, one that likely involves shared teams and resources across projects.

For most projects, in addition to this technical staff and the project leader, you also need another business-side role filled: an *operational liaison* (a.k.a. *analytics translator, business translator, data product partner*, or *innovation marshal*). This person bridges the gap between tech and business, between the technical project staff and the line-of-business stakeholders in charge of the operations that the project will alter. The operational liaison ensures the model will be understood and embraced by those running operations. This person works within—or closely with—the line-of-business team. The pertinent team may be in marketing, website operations, fraud investigations, or financial credit application processing. The liaison is involved from the get-go, delivering feedback to the ML team to ensure the prediction goal, deployment plan, and performance objectives are aligned with the operational team's needs.

Sourcing the ML Project Team

Where do you get these people? After all, folks with these very particular skills can be hard to come by.

Well, when you need something, there are three ways to get it: Buy it, rent it, or make it. For a staff member, that means hire someone new, engage consultants, or train existing staff.

External consultants are often central when you're launching a pilot ML project, since, that way, you don't need to commit to hiring ML experts until you have more firmly established the value of the ML application that you're pursuing for the first time. Outsourcing to consultants can be expensive, but the good news is that you can often keep this expense reasonably low, since you need these experts only for a relatively light engagement. During the first three planning steps of bizML, a consultant can be engaged for light "consulting" in the literal sense of the word, helping to refine the project plan down to the details of the prediction goal, and informing the data requirements. Then, most of step 4, data preparation, can be handled by your existing internal staff. The core predictive modeling is where you'll need the most intensive expert assistance, but that step is relatively short in terms of person hours and calendar days.

Eventually, as you tackle more ML projects and move toward establishing more in-house expertise, training your existing staff is a favored way to "grow your own" ML team. In addition to training your staff, ML leadership innovator and professor Bryan Bennett advocates for a complementary approach he calls the *DataScienceStein* approach—à la Frankenstein's monster. Since data scientists are hard to come by and data science involves so many varied skills, he suggests "building your data scientist out of a team of people currently on staff or readily available in the marketplace" with the right complementary skills.

ML experts come from all kinds of backgrounds and scientific fields. While your staff further develops its ML skills, keep in mind that we see ML practitioners who have moved laterally from all walks of life, including neurosurgeons, physicists, and psychologists. It turns out that experience in all kinds of quantitative fields often translates quite nicely to working in ML.

As you bring staff onboard or upskill them, beware the overzealous notion of an "all-capable" data scientist. It's better your staff develop the specific skills that your ML project needs than for them to pursue the overly ambitious goal of becoming an individual super-employee capable of performing any and all data-related tasks. Make sure the project roles are well defined and the team member skills meet the needs of each role.

Regardless of how you source your staff for an ML project, they won't come cheap. For how much time will you need to utilize these in-demand experts?

Projecting the Project: How Long It Will Take

> In theory there is no difference between theory and practice, while in practice there is.
>
> —Benjamin Brewster, nineteenth-century American industrialist

You might be in for a long journey. ML projects vary in duration as widely as enterprise projects in general. And they suffer delays as often as any kind of operational change.

In the very best of scenarios, the stars could align for a two-month project. This could be the case, for example, if you're cranking out a model to target an existing direct marketing operation—swapping it into an established process, and only requiring batch deployment from within the ML software without the need to export the model.

In contrast, Jack's project at UPS took years to launch at full scale. It was a first-of-its-kind initiative at the company, deploying a model within a complex, semi-automatic system for which extensive in-person staff training was required across many sites.

Only the people and protocols at your company can properly project the timeline. But, to help establish your rough estimate, here are some ballpark ranges for each step, with the caveat that, for each one, the sky could turn out to be the limit:

Steps 1, 2, 3: Establish the deployment goal, prediction goal, and evaluation metrics: two weeks to three months.

Step 4: Prepare the data: four weeks to five months.

Step 5: Train the model: three weeks to two months.

Step 6: Deploy the model: three weeks to one year.

Even with great excitement about a new ML initiative, the first three pre-production steps make for an era of planning, socializing, and greenlighting. Your up-front pitch may take less than two minutes, but you will need to ride that elevator quite a few times.

Of the three technical steps, the most sophisticated science takes the least calendar time. Data preparation is a perpetually underestimated bottleneck, and model deployment can be as well, especially if it requires the model to be integrated into existing real-time systems. On the other hand, although model-training demands the greatest depth of ML experience and expertise, it's a relatively contained, isolated process—the steps just before and after it effectively buffer the core number crunching itself from many enterprise complexities.

Another consideration makes the project timeline harder to estimate and potentially drags it out: Following the six steps is not a linear process.

Backtracking: Iteratively Looping on the Steps

Almost always what happens in my most successful engagements is you get trust and buy-in up front, you try to solve the problem, and then you're wrong about, like, half of the assumptions . . . you uncover things you have to adjust and adapt in the middle of the project.

—Dean Abbott, renowned consultant and chief data scientist of Abbott Analytics

With bizML, you backtrack a lot, looping back to a previous step as new insights come to bear. At each iteration, the team must reconvene with stakeholders to revisit earlier choices. Here are some examples:

- While preparing the data, data scientists or data engineers discover that there aren't enough positive examples available. This triggers team members to reconvene and modify the prediction goal to be one for which there are plenty.

- While training the model, data scientists detect a bug in the training data: a data leak. This circles back to data preparation.

- After training the model, its performance is disappointing and decision makers say it's not ready for deployment. This can mean circling back to any previous step, such as reconsidering the prediction goal, reconsidering the evaluation metrics, or enhancing the training data.

- During deployment, field tests show that model-scoring is too slow, because the model itself is too complex. By circling back to model training, it may be possible to generate a simpler, faster model that exhibits only a minimal loss in predictive performance. In one famous case, Netflix decided not to deploy the complex model with which competitors won the firm's $1 million contest to improve movie recommendations. Netflix elected to deploy internally developed models instead.

- After deployment, the business context changes—new regulations emerge or new strategic imperatives are handed down that prompt a change to the deployment goal. Although this means beginning again with step 1, much of the work completed throughout all the bizML steps may be repurposable during this new project iteration.

In the best of cases, you circle back due to a happy surprise or a new inspiration. For example:

- During model training, an input variable unexpectedly proves to be important, inspiring the introduction of related data sources during data preparation.

Critically, each time the project backtracks, a cross-disciplinary team jointly navigates, incorporating the insights of data scientists as well as those of business stakeholders and operations managers.

Life after Launch: Maintaining the Model

"Developing and deploying ML systems is relatively fast and cheap, but maintaining them over time is difficult and expensive," warns a technical paper from Google. This writing has gained some notoriety among data scientists—although perhaps not enough, in my opinion. The paper's title would strike fear in the most stoic accountant: "Hidden Technical Debt in Machine Learning Systems."

When you launch astronauts into space, you commit yourself to a new job: You've got to keep them alive. Likewise, once it's in play, sustaining a model's viability moving forward takes maintenance, monitoring, and vigilance. The model and the deployment infrastructure that cradles it join the ranks of mission-critical enterprise systems. These things require upkeep.

For one thing, models stagnate. If they remain unchanged, they degrade. The world changes around them. The economy shifts and customer behavior patterns evolve. As a result, the data over which a model was trained becomes less pertinent, less representative of today's world. After all, that training data becomes a part of the more distant past every day. Over time, your model inevitably devolves into a defunct dinosaur, a phenomenon known as *model drift*. This prompts the need to monitor model performance over time. The tools and techniques for doing so are sometimes collectively called *ML observability*.

The remedy is to periodically update the model. That's standard protocol. This normally means training a whole new model over more recent data (incrementally updating an existing model is an alternative, but only rarely worth the complexity involved). For some projects, the refresh is daily and for others it's annual. It can be triggered when model performance weakens or can be scheduled at regular intervals. When the world changes drastically, due to political upheaval, natural disasters, or a pandemic, models become outdated more quickly. Such events potentially highlight ML's strength: the capacity to adapt to a changed world. But this capacity is realized only if you refresh the model on data that reflects the new world in which you now live.

It's worth clarifying that, in a certain sense, some deployed models adapt on their own between updates in that the inputs are always kept updated. For example, FICO Falcon's fraud detection model is updated only once a year for the 9,000-plus banks that use it. But some of its inputs are engineered to track the ever-changing usage patterns of each individual cardholder. For example, if a cardholder begins to regularly shop at small online stores, an input could reflect this change in the individual's tendencies, showing such purchases as less anomalous for them. This level of adaptation is continuous and ongoing between model updates. What changes with periodic model updates is how the model weighs and considers such an input, whereas the input itself is continually updated.

The expense of upkeep only adds to the already substantial price tag attached to each ML project. The resources needed just to get to deployment include staff, software, and deployment infrastructure, such as data pipelines and possibly an upgraded operational system to incorporate model scoring. The total that these expenditures come to can grow almost as much as for any kind of business initiative.

But ML's value eases the sticker shock. Divide the estimated win by the cost and you have your potential ROI. It's bound to be strong for one simple reason: Prediction pays.

But beyond the business gains, there's a nonfinancial consideration to also account for.

Morality Matters

> AI is a superpower that enables a small team to affect a huge num-
> ber of people's lives . . . make sure the work you do leaves society
> better off.
>
> —Andrew Ng

When you use ML, you aren't just optimizing models and streamlining business. You're governing. In effect, models embody and implement policies that control access to opportunities and resources, such
as credit, employment, housing—and even freedom, when it comes to
arrest-prediction models that inform parole and sentencing. Insurance
risk models determine what each policyholder must pay, and targeted
marketing determines who gains discounts, exclusive deals, and even
the awareness of certain financial products.

When ML acts as the gatekeeper to these opportunities, it can perpetuate or magnify social injustice, adversely affecting underprivileged
groups by undeservingly denying access disproportionately often. Here
are four ways in which that can happen, among others:

1. *Discriminatory models*: Models that take a protected class such as race
   or national origin as an input so that their decisions are directly based
   in part on that class. These models discriminate explicitly, doing so
   more visibly and detectably than a person who discriminates but
   keeps private the basis for their decisions. For example, such a model
   could penalize a Black person for being Black. Although outlawed
   in some contexts and relatively uncommon so far, some decorated
   experts in ML ethics loudly advocate for allowing protected classes as
   model inputs.

2. *Machine bias*: Unequal false-positive rates between groups, which
   means the model incorrectly denies approval for or access to opportunities to one group more often than another. This can and often
   does occur even if the model is not *discriminatory* (per above), since
   a model can employ other, unprotected input variables as proxies

for a protected class. For example, ProPublica famously exposed a rearrest-prediction model that wrongly jails Black defendants more often than White defendants.

3. *The coded gaze*: When a group is underrepresented in the training data, the resulting model won't work as well for members of that group. This results in exclusionary experiences, such as when a facial recognition system fails for Black people more often than for people of other races. Also known as *representation bias*, this phenomenon can also occur for speech recognition.

4. *Inferring sensitive attributes*: A model's predictions can reveal group membership, such as sexual orientation, whether someone is pregnant, whether they'll quit their job, or whether they're going to die. Researchers have shown that it is possible to predict race based on Facebook likes, and officials in China use facial recognition to identify and track the Uighurs, a minority ethnic group systematically oppressed by the government. In these cases, sensitive information about an individual is derived from otherwise innocuous data.

The question to always ask is, "For whom will this fail?" says Cathy O'Neil, author of *Weapons of Math Destruction* and one of the most visible activists in ML ethics. This fundamental question conjures the four issues above and many others as well. It's an ardent call to action that reminds us to pursue ethical considerations as an exercise in empathy.

Only proactive leaders can meet these ethical challenges. Companies using ML are mostly frozen by the cosmetics demanded by corporate public relations. It's often only to posture when firms call for ML deployment to be "fair, unbiased, accountable, and responsible." These are vague platitudes that don't alone guide concrete action. Declaring them, corporations perform *ethics theater*, protecting their public image rather than protecting the public. Rarely will you hear a firm come down explicitly on one side or the other for any of the four issues I listed above, for example.

O'Neil has taken on the indifference to these and other issues with another weapon: shame. She advocates for shaming as a means to battle corporations that deploy analytics irresponsibly. Her more recent book, *The Shame Machine*, takes on "predatory corporations" while criticizing shame that punches down rather than up. The fear of shame delivers clients for her model-auditing consulting practice. "People hire me to look into their algorithms," says O'Neil. "Usually, to be honest, the reason they do that is because they got in trouble, because they're embarrassed . . . or sometimes it's like, 'We don't want to be accused of that and we think that this is high-risk.'"

But I would invite you to also consider a higher ideal: Do good rather than avoid bad. Instead of dodging shame, make efforts to improve equality. Take on the setting of ethical ML standards as a form of social activism. To this end, define standards that take a stand rather than only conveying vague platitudes. For starters, I advocate for the following standards, which I consider necessary but not sufficient: Prohibit discriminatory models, balance the false-positive rates across protected groups, deliver on a person's right to explanation for algorithmic decisions—at least in the public sector—and diversify analytics teams.

Your role is critical. As someone involved in initiatives to deploy ML, you have a powerful, influential voice—one that is quite possibly much more potent than you realize. You are one of a relatively small number who will mold and set the trajectory for systems that automatically dictate the rights and resources to which great numbers of consumers and citizens gain access. Allan Sammy, director of data science and audit analytics at Canada Post, put it this way: "A decision made by an organization's analytic model is a decision made by that entity's senior management team."

ML can help rather than hurt. Its widening adoption provides an unprecedented new opportunity to actively fight injustice rather than perpetuate it. When a model shows the potential to disproportionately affect a protected group adversely, it has put the issue on the table and under a spotlight by quantifying it. The analytics then provide quantitative options to tackle injustice by adjusting for it. And the very same operational framework to automate or support decisions with ML can be leveraged to deploy models adjusted to improve social justice.

As you follow this book's practice to get ML successfully deployed, make sure you're putting this powerful technology to good use. If you optimize only for a single objective such as improved profit, there will be fallout and dire ramifications. But if you adopt humanistic objectives as well, science can help you achieve them. O'Neil sees this, too: "Theoretically, we could make things more fair. We could choose values that we aspire to and embed them in code. We could do that. That's the most exciting thing, I think, about the future of data science."

Over the last decade, I have spent a considerable portion of my work on ML ethics. For a more in-depth dive, such as a visual explanation of machine bias, a call against models that explicitly discriminate, and more details regarding the standards I propose, see my writing and videos at

www.civilrightsdata.com.

Acknowledgments

My wife, Luba Gloukhova, contributed immensely to this book. While any author's partner is tasked with cultivating patience and perhaps taking on extra parenting responsibilities, Luba carried the additional burden of being a subject matter expert herself: She's a data scientist, and an insightful, experienced one at that (see her consultancy at www .datagie.com). As such, I endlessly solicited her counsel at meals, on walks, and in playgrounds. She served as a 24/7 sounding board as well as an intensive reviewer. Luba's more formal partnerships with me also provided value for this book, including her work as the founding chair of the Deep Learning World conference series, the editor-in-chief of the *Machine Learning Times*, and the content editor of my online course. I could not have written this book without Luba's enthusiasm and support. I must add, though, that her virtuoso performance as loving wife and mother transcends all of that!

My parents and in-laws also went to extra lengths to support me in this project. Lisa Schamberg (mother), Andrew Siegel (father), Anna Gloukhov (mother-in-law), and Maya Kanyuka (grandmother-in-law) contributed priceless encouragement as well as insights for my writing.

My literary agent, the incomparable Jim Levine of Levine Greenberg Rostan, provided critical corrections to my course during early, formative stages of this project and he followed through in later stages to help me bring the book's vision into a more carefully crafted reality. If not for his keen wisdom and business acumen, this book would have made a lot less sense.

My editor, Catherine Woods of the MIT Press, knows what makes a book readable, relatable, and relevant. Her feedback was critical as I worked through many balancing acts in this book's formulation. And Abbie Lundberg, despite her responsibilities as editor in chief of *MIT Sloan Management Review*, took the time to dig in deep and provide much-needed feedback that improved the book greatly. Earlier on, Emily Taber believed in this project and took me on as an author. Despite being on her way out by the time I finished the first draft—to another publisher after thirteen years at the MIT Press—she went the extra mile providing in-depth feedback.

Laurie Harper, Alison Jones, David Lamb, Barbara Monteiro, Barbara Oakley, and Myles Thompson served as mentors with a great deal of experience in the publishing industry. They advised me from this project's early conception on navigating the publication process, as well as providing substantive input on the project's formulation.

A career in tech is incomplete without a group of chums with whom you not only collaborate but, more importantly, talk shop on the regular. Each a senior consultant as well as an entrepreneur, Dean Abbott, John Elder, Karl Rexer, and James Taylor have enhanced my knowledge and enjoyment over the years and—more recently—provided extensive reviewing of an earlier draft of this book.

I've also learned tons from a much broader swathe of machine learning professionals: the 18,000 attendees and speakers who have participated at the conference series I founded in 2009, Machine Learning Week (www.machinelearningweek.com), formerly Predictive Analytics World. Because of the productive and vibrant community that has attended our dozens of events internationally, I've enjoyed a 360-degree view of the industry that greatly informed this book. I'd like to thank my business partner in its production, Matthew Finlay, and his crackerjack team at Rising Media, who know how to make events excite and unite.

Many at the University of Virginia Darden School of Business— where I held a one-year position as Bodily Bicentennial Professor of Analytics during the writing of this book—contributed significantly to my formulation of ideas as I honed the methodology described herein. I am grateful to Samuel Bodily, the professor emeritus after whom my position was named, who personally provided a great deal of input and guidance on my work, much of which made its way into this book. Various faculty at Darden and other UVA departments also provided such input, including Michael Albert, Alexander Cowan, Renée Cummings, Rupert Freeman, Yael Grushka-Cockayne, Marc Ruggiano, Bill Scherer, Eric Tassone, and Sasa Zorc.

I would like to thank Jack Levis, Scott Zoldi, and Gerhard Pilcher, all innovators in ML deployment who endured my extensive interviews filled with many probing questions.

It takes a village. It was my great fortune that the following people invested time and care reviewing earlier drafts of these chapters: Rich Heimann, Eugene Kirpichov, Sam Koslowsky, Barry Lyons, Matthew Mayo, Glenn McMahan, Gregory Piatetsky-Shapiro, Steven Ramirez, Jonathan Sloan, Graham Southorn, David Stephenson, Martin Szugat, Morgan Vawter (in addition to providing this book's inspirational foreword), and Evan Wimpey.

Thank you, Sean O'Brien and Catherine Truxillo at SAS for their support in producing and hosting my online course series, "Machine Learning Leadership and Practice: End-to-End Mastery" (www.machinelearning .courses), the content of which served as a foundation for this book. Thanks also to William Goodrum for assistance with the course's content.

Many thanks to *Harvard Data Science Review* for originally publishing my writing about UPS's deployment story in an article entitled "To Deploy Machine Learning, You Must Manage Operational Change— Here Is How UPS Got It Right" (vol. 5, no. 2, Spring 2023). The article tells the part of the UPS story covered mostly by this book's sixth chapter, as well as some aspects covered in the introduction.

Thank you to *Scientific American*, which originally published my article on the *accuracy fallacy*, an earlier version of chapter 3's opening.

Thanks to the supremely gifted designer Matt Kornhaas for the figures throughout these chapters and the extraordinary artist Daneen Wilkerson for the slingshot image on the book's cover.

Finally, here's a shout-out to the extra-special educators, of whom I had more than my fair share: Thomas McKean (kindergarten), Chip Porter (grades 4–6), Margaret O'Brien (Burlington High School, Vermont), Harry Mairson (Brandeis University), Richard Alterman (Brandeis University), James Pustejovsky (Brandeis University), and Kathleen McKeown (Columbia University).

The aforementioned have molded me and bolstered this book. Nevertheless, I alone take responsibility for errors or failings of any kind in its contents.

About the Author

Eric Siegel, PhD, is a leading consultant and former Columbia University professor who helps companies deploy machine learning. He is the founder of the long-running Machine Learning Week conference series, the instructor of the acclaimed online course "Machine Learning Leadership and Practice—End-to-End Mastery," executive editor of the *Machine Learning Times*, and a frequent keynote speaker. He wrote the bestselling *Predictive Analytics: The Power to Predict Who Will Click, Buy, Lie, or Die*, which has been used in courses at hundreds of universities. Eric's interdisciplinary work bridges the stubborn technology/business gap. At Columbia, he won the Distinguished Faculty award when teaching the graduate *computer science* courses in ML and AI. Later, he served as a *business school* professor at UVA Darden. Eric also publishes op-eds on analytics and social justice.

Eric has appeared on Bloomberg TV and Radio, BNN (Canada), Israel National Radio, National Geographic Breakthrough, NPR Marketplace, Radio National (Australia), and TheStreet. Eric and his previous book have been featured in *Big Think*, *Businessweek*, *CBS MoneyWatch*, *Contagious Magazine*, the *European Business Review*, the *Financial Times*, *Forbes*, *Fortune*, *GQ*, *Harvard Business Review*, *Huffington Post*, the *New York Review of Books*, the *New York Times*, *Newsweek*, *Quartz*, *Salon*, the *San Francisco Chronicle*, *Scientific American*, the *Seattle Post-Intelligencer*, the *Wall Street Journal*, the *Washington Post*, and *WSJ MarketWatch*.

» Eric Siegel is available for select lectures. To inquire: www.MachineLearning Keynote.com

» Attend the conference founded by the author: www.MachineLearningWeek .com

» Access the author's online course: www.MachineLearning.courses

» Follow the author: @predictanalytic or www.linkedin.com/in/predictiveanalytics

Index